Geothermal Energy

Edited by Basel I. Ismail

Published in London, United Kingdom

IntechOpen

Supporting open minds since 2005

Geothermal Energy
http://dx.doi.org/10.5772/intechopen.87837
Edited by Basel I. Ismail

Contributors
Mostafa Khosravy, Arif Widiatmojo, Isao Takashima, Youhei Uchida, Mitsuo Matsumoto, Mébrouk Benziada, Ziyodulla Yusupov, Mohamed Almaktar, Chiranjit Maji, Hirok Chaudhuri, Saroj Khutia, Basel I. Ismail

First published in London, United Kingdom, 2022 by IntechOpen
IntechOpen is the global imprint of INTECHOPEN LIMITED, registered in England and Wales, registration number: 11086078, 5 Princes Gate Court, London, SW7 2QJ, United Kingdom
Printed in Croatia

British Library Cataloguing-in-Publication Data
A catalogue record for this book is available from the British Library

Additional hard and PDF copies can be obtained from orders@intechopen.com

Geothermal Energy
Edited by Basel I. Ismail
p. cm.
Print ISBN 978-1-83969-233-8
Online ISBN 978-1-83969-234-5
eBook (PDF) ISBN 978-1-83969-235-2

For EU product safety concerns:
IN TECH d.o.o., Prolaz Marije Krucifikse Kozulić 3, 51000 Rijeka, Croatia, info@intechopen.com or visit our website at intechopen.com.

Meet the editor

Dr. Basel Ismail is currently an associate professor and chair of the Department of Mechanical Engineering, Lakehead University, Thunder Bay, Ontario, Canada. In 2004, Prof. Ismail earned his Ph.D. in Mechanical Engineering from McMaster University, Hamilton, Ontario, Canada. From 2004 to 2005, he worked as a postdoctoral researcher in the Engineering Physics Department at McMaster University. His specialty is in engineering heat transfer, thermodynamics, energy conversion and storage, CO_2 membrane gas separation for greenhouse gas (GHG) reduction, and renewable energy technologies. Dr. Ismail's research activities are theoretical and applied in nature. Currently, his research interests are focused on green engineering technologies related to alternative and renewable energy systems for power generation, heating, and cooling. Dr. Ismail was the leading research investigator in a collaborative project with Goldcorp-Musselwhite Canada Ltd. and Engineering, Lakehead University from 2007 to 2010. This innovative project was state-of-the-art in geothermal heat pump technology applied in Northwestern Ontario, Canada. Dr. Ismail has edited four books and published many technical reports, chapters, and articles related to his research areas in reputable international journals and conferences.

Contents

Preface

The term "geothermal energy" refers to the Earth's natural heat energy. The continual heat energy flux coming from the Earth's core to the surface is the source of geothermal energy. The Earth's geothermal resources are huge; for example, the portion of geothermal energy stored at a depth of 3 km is estimated to be 1,194,444,444 TWh, which is substantially more than the total energy equivalent of all fossil fuel resources combined, which is estimated to be 1,010,361 TWh. In general, geothermal energy can be utilized for power generation or direct heating applications. It is regarded as an environmentally beneficial clean energy source that, when used to generate electrical power, has the potential to considerably reduce greenhouse gas (GHG) emissions. It is reported that global net electrical power demand will increase by nearly 85% between 2004 and 2030 to 30,364 TWh in 2030 from 16,424 TWh in 2004, making the use of geothermal energy for electricity generation an appealing solution, especially with advances being made in innovative technological methods of drilling and power generation schemes.

This book covers various interesting topics of research related to geothermal energy due to its significant utilization and prospects. It is the result of contributions from several researchers and experts worldwide. It is hoped that the book will be a useful source of information and basis for extended research for researchers, academics, policy makers, and practitioners in geothermal energy.

This book contains seven chapters. Chapter 1 introduces some fundamental aspects of Organic Rankine Cycle (ORC) binary-fluid power technology using low-temperature geothermal energy resources with a detailed numerical example as an illustration of its thermodynamic performance. Chapter 2 presents the geothermal energy resource in terms of the types of power plants, principle of electricity generation, and current world status of geothermal resource utilization. Chapter 3 discusses a simple technique by means of the exploration study at a field site in Eastern India. It estimates the potential of geothermal energy generated inside the Bakreswar reservoir. Chapter 4 describes the state of the art of several existing low-temperature district heating systems (LTDHs). The advantages of LTDH networks over the traditional district heating networks were discussed. Reviewed cases and studies intensified the energy efficiency potential of LTDH. This chapter suggests that this system provides a unique opportunity to integrate renewable heat sources such as geothermal and solar as much as possible.

Chapter 5 describes an approach to estimate a geothermal reservoir's productivity during active exploration and development of a geothermal prospect. This approach allows a reservoir model to be updated by overcoming the severe time limitations associated with accessing sites for drilling and well testing under snowy and mountainous conditions. Chapter 6 discusses the results of hydrogeochemical prospecting of hot springs in central northern Algeria. Finally, Chapter 7 discusses the numerical evaluation of thermal response test results in Bangkok, Thailand, and Hanoi, Vietnam. In this chapter, a moving infinite line source analytical model to evaluate the value of thermal conductivity and groundwater flow velocity is applied. Furthermore, the five-year performance of the ground source heat pump system coupled with two vertical ground heat exchangers in Bangkok and Hanoi are evaluated.

I would like to thank all chapter authors for their efforts. I would also like to thank Author Service Manager Ms. Romina Rovan at IntechOpen for her excellent efforts in managing the publication of this book.

Dr. Basel I. Ismail, P.Eng.
Associate Professors and Chair,
Faculty of Engineering,
Department of Mechanical Engineering,
Lakehead University,
Thunder Bay, Ontario, Canada

Chapter 1

Introductory Chapter: ORC Power Generation Technology Using Low-Temperature Geothermal Energy Resources: A Conceptual Case Study

Basel I. Ismail

1. Introduction

The term "Geothermal Energy" refers to the Earth's natural heat energy. The continual heat energy flux coming from the Earth's core to the surface is the source of geothermal energy. The Earth's geothermal resources are huge; for example, the portion of geothermal energy stored at a depth of 3 km is estimated to be 1,194,444,444 TWh, which is substantially more than the total energy equivalent of all fossil fuel resources combined, which is estimated to be 1,010,361 TWh [1]. Geothermal energy is regarded as an environmentally beneficial clean energy source that, when used to generate electrical power, has the potential to considerably reduce GHG emissions. It is reported that global net electrical power demand will rise by nearly 85% between 2004 and 2030, increasing to 30,364 TWh in 2030 from 16,424 TWh in 2004, making the use of geothermal energy for electricity generation an appealing solution, especially with advances being made in innovative technological methods of drilling and power generation schemes. Geothermal energy resources differ regionally based on the temperature and depth of the resource, the availability of ground water, and the chemical composition of the rock [2]. It is distinct from other conventional and renewable energy sources in that it is always accessible, steady throughout the year, regardless of weather conditions, and has an inherent storage potential. The temperature of geothermal energy resources typically ranges approximately from 50 to 350°C. Geothermal resources near volcanic regions and island chains tend to have a high resource temperature with temperature typically greater than 200°C. Medium-temperature, ranging from 150 to 200°C, and low-temperature geothermal resources of less than 150°C are typically found widely in most continental regions and considered to be the most commonly available geothermal energy resources [3]. The geothermal binary cycle technology known as the Organic Rankine Cycle (ORC) technology can successfully generate power from medium- and low-temperature geothermal energy (LTGE) resources. LTGE-ORC technology emits almost no greenhouse gases into the environment and is an appealing technology because of its simplicity and small number of components, all of which are common and commercially accessible. More information related to a number of past and existing successful ORC binary power plants can be found in Refs. [1–4]. In this introductory chapter, the fundamental concept of ORC binary fluid power technology using LTGE geothermal resources is introduced with a detailed numerical example, as an illustration of its thermodynamic performance.

2. Fundamental concept of a binary fluid ORC system using LTGE resources

Figure 1 depicts a schematic representation of an ORC binary fluid system utilized for electric power generation using LTGE resource.

The initial (main) fluid collected from the LTGE resource *via* the production well is the geo-fluid. The geo-fluid transports heat from the liquid-dominated LTGE resource (being the heat carrier) and effectively transfers this heat to the low-boiling point organic-based working fluid (the secondary fluid) through an efficient heat exchanger. Typical ORC organic fluids may include pure hydrocarbons (e.g., pentane, butane, propane), refrigerants (e.g., R134a, R218, R123, R113, R125), or organic mixtures. More details about the selection criteria of these ORC organic fluids for optimal performance can be found in Refs. [1–4]. The ORC is a thermodynamic Rankine cycle that uses the organic working fluid instead of steam (water). In this binary fluid ORC LTGE system, the low-boiling point organic liquid absorbs the heat which is transferred by the geo-fluid and boils at a relatively much lower temperature (compared to water) and as a result develops significant vapor pressure sufficient to drive the axial-flow or radial-inflow turbine. The turbine is coupled to an electric generator, which converts the turbine mechanical shaft power into electrical power. The organic working fluid expands across the turbine and then is cooled and condensed in the condenser before it is pumped back as a saturated liquid to the heat exchanger using a condensate pump to be re-evaporated, and the power cycle repeats itself.

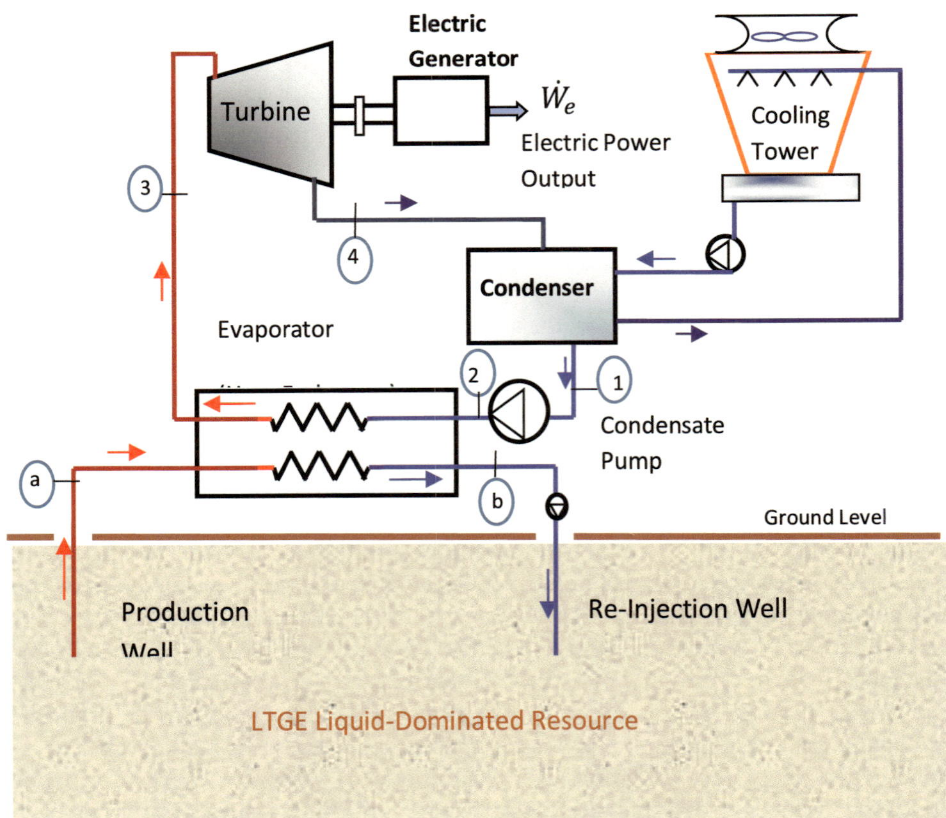

Figure 1.
A schematic showing the fundamental concept of an ORC binary fluid technology utilizing LTGE resource for power generation.

3. Thermodynamic analysis of an ORC system utilizing a low-temperature geothermal energy resource: *A case study with numerical illustration*

A small-scale modular binary fluid ORC geothermal power generation system is proposed for design and installation in Vancouver, BC Canada, near a site characterized by a low-temperature geothermal energy resource. The production well in this resource is capable of supplying hot geo-fluid (mainly liquid water) at a temperature of 90°C (see **state a, Figure 1**). The proposed ORC system utilizes R-134a as a working fluid. In this ORC system shown in **Figure 1**, it is required to provide R-134a at a mass flow rate $\dot{m}_R$of 6.25 kg/s as saturated vapor at 85°C to an inflow radial turbine (**state 3**). The ORC condenser operates at constant pressure with a constant phase-change temperature of 40°C. R-134 then enters an ideal ORC-pump as a saturated liquid (**state 1, Figure 1**). The density of the geo-fluid is assumed to be constant at approximately 1000 kg/m^3. The geo-fluid liquid exits the ORC-evaporator (**state b, Figure 1**) to be re-injected into the geothermal resource at 35°C with constant specific heat capacity $C_{p,Geo}$ = 4.185 kJ/kg.°C.

Thermal design constraints: The ORC-evaporator (counter flow-HEx) effectiveness $\varepsilon_{Evap,ORC}$= 90%; the turbine isentropic efficiency η_T = 100%; negligible pressure drops in the ORC-piping systems.

Electric generator specification: η_{EG} = 92%.

Electricity-driven pump efficiency = 100%.

Required: For this conceptual ORC power generation system, the following thermodynamic performance indicators are determined:

a. The R-134a-pumping power $\dot{W}_{Pump}$ (kW) requirement for the ORC system.

b. The heat transfer rate input $\dot{Q}_{Evap,ORC}$ (kW) to the ORC system.

c. The turbine supplied power $\dot{W}_{Turb}$ (kW).

d. The electric generator power output $\dot{W}_{EG}$ (kW).

e. The net power $\dot{W}_{net,ORC}$ (kW) delivered by the ORC system.

f. The thermal efficiency $\eta_{th,ORC}$ (%) of the ORC system.

g. The specific heat transfer q_{Geo} (kJ/kg) supplied by the low-temperature geothermal resource to the ORC system.

h. The heat transfer rate $\dot{Q}_{Geo}$(kW) supplied by the LTGE resource to the ORC system.

i. The mass flow rate of the geo-fluid $\dot{m}_{Geo}$(kg/s) to be extracted from the low-temperature geothermal resource.

j. The volumetric flow rate of the geo-fluid $\dot{V}_{Geo}$(L/min) to be extracted from the geothermal resource.

k. The overall energy conversion efficiency η_{Geo} (%) of the low-temperature geothermal ORC power generation system.

4. Methodology, analysis, and results

Application of the steady-state energy balance, mass balance, and other thermodynamic relationships over the ORC pump (stream 1–2), turbine (stream 3–4), evaporator (stream 2–3), electric generator (EG), and the geothermal resource (stream a-b), yields the following set of model equations (**Tables 1** and **2**):

$$\dot{W}_{pump} = \dot{m}_R(h_2 - h_1) \tag{1}$$

$$w_{pump} = h_2 - h_1 \tag{2}$$

$$h_2 - h_1 = v_1(p_2 - p_1) \tag{3}$$

$$\dot{Q}_{Evap,ORC} = \dot{m}_R(h_3 - h_2) \tag{4}$$

$$\dot{W}_{Turb} = \dot{m}_R(h_3 - h_4) \tag{5}$$

$$\dot{W}_{EG} = \eta_{EG}\dot{W}_T \tag{6}$$

$$\dot{W}_{net,ORC} = \dot{W}_{EG} - \dot{W}_{pump} \tag{7}$$

$$\eta_{th,ORC} = \frac{\dot{W}_{net,ORC}}{\dot{Q}_{Evap,ORC}} \tag{8}$$

$$\dot{Q}_{Geo} = \dot{m}_{Geo}q_{Geo} \tag{9}$$

$$\dot{m}_{Geo} = \frac{\dot{Q}_{Geo}}{q_{Geo}} \tag{10}$$

$$\dot{V}_{Geo} = \frac{\dot{m}_{Geo}}{\rho_{Geo}} \tag{11}$$

$$\dot{Q}_{Geo} = \dot{m}_{Geo}(h_a - h_b) \tag{12}$$

$$q_{Geo} = c_{p,Geo}(T_a - T_b) \tag{13}$$

$$\dot{Q}_{Geo} = \frac{\dot{Q}_{Evap,ORC}}{\varepsilon_{Evap,ORC}} \tag{14}$$

$$\eta_{Geo} = \frac{\dot{W}_{net,ORC}}{\dot{Q}_{Geo}} \tag{15}$$

State #	T (°C)	P (kPa)	v (m^3/kg)	h (kJ/kg)	s (kJ/kg.°C)	x (%)
1	40	1017.0	0.000873	256.54		0.0
2		2926.2		258.21		
3	85	2926.2		428.10		100
4				409.14	1.6782	93.5
a	90					
b	35					

Table 1.
Thermodynamic properties of R-134a at the given conditions.

Required	Calculated numerical value
$\dot{W}_{Pump}$ (kW)	10.42
$\dot{Q}_{Evap,ORC}$ (kW)	1061.33
$\dot{W}_{Turb}$ (kW)	118.50
$\dot{W}_{EG}$ (kW)	109.02
$\dot{W}_{net,ORC}$ (kW)	98.60
$\eta_{th,ORC}$ (%)	9.3
q_{Geo} (kJ/kg)	230.18
$\dot{Q}_{Geo}$ (kW)	1179.31
$\dot{m}_{Geo}$ (kg/s)	5.13
$\dot{V}_{Geo}$ (L/min)	307.5
η_{Geo} (%)	8.4

Table 2.
Summary of the ORC-LTGE performance results.

Author details

Basel I. Ismail
Department of Mechanical Engineering, Lakehead University, Canada

*Address all correspondence to: baseliai@gmail.com

References

[1] Ismail BI. Chap 18: Power generation using nonconventional renewable geothermal & alternative clean energy technologies. In: Carayannis EG, editor. Planet Earth 2011 – Global Warming Challenges and Opportunities for Policy and Practice. London, UK: InTech Open Access Publishing Company; 2011

[2] Ismail BI. Advanced electrical power generation technology using renewable & clean low-enthalpy geothermal energy sources. Recent Patents on Mechanical Engineering. 2011;**4**(2): 168-179

[3] Ismail BI. Advances in Geothermal Energy. Rijeka, Croatia: InTech Open Access Publishing Company, 2016. 174 pages. DOI: 10.5772/60623

[4] Ismail BI. Chapter 1: Power generation using geothermal low-enthalpy resources and ORC technology. In: Ismail BI, editor. Renewable Geothermal Energy Explorations. London, UK: IntechOpen; 2019. DOI: 10.5772/intechopen.84390. Available from: https://www.intechopen.com/chapters/66034

Chapter 2

Geothermal Power Generation

Ziyodulla Yusupov and Mohamed Almaktar

Abstract

Bulk power system based on fossil fuels becomes less reliable and stable in economic terms, technically more labor-consuming and harmful environmental impact. These problems have led many countries to find ways to supply the electricity from a green and sustainable energy source. The electricity derived from renewable energy sources such as hydro, solar, wind, biomass and geothermal refers to as green and sustainable energy. Geothermal energy is not only utilized for electric power generation, but it is also exploited to generate environmentally friendly heat energy. As of the end of 2018, geothermal global cumulative installed capacity exceeded 13 GW, generated an energy of about 630 peta joule (PJ). This chapter presents the geothermal energy resource in terms of the types of power plants, principle of the electricity generation and current world status of geothermal resource utilization. The issues such as advantages and disadvantages of geothermal energy economically and environmentally and means to overcome shortcomings are also considered. The main barriers for the development of geothermal industry include high resource and exploration risk, overall high development cost particularly drilling, and inadequate financing and grant support. The global averaged cost of electricity for the geothermal facility is nearly 0.072 USD/kWh as compared to 0.056 for onshore wind and 0.047 USD/kWh for hydropower. However, the technology is rather competitive to other renewables such as concentrating solar power (0.185 USD/kWh) and offshore wind (0.127 USD/kWh). Meanwhile, further research and development is critically needed to eliminate the non-condensable gases (NCGs) associated with the geothermal power generation.

Keywords: geothermal energy, geothermal power plants, geothermal electricity, economic and environmental impact, geothermal worldwide status

1. Introduction

Geothermal energy is heat energy from Earth's interior which generated from radioactive breakdown and frequently recurring heat losses from Earth's formation. The Earth's heat capacity is approximately 1×10^{19} TJ or 2.8×10^{15} TWh [1]. The Earth's heat conduction is 44.2 TW [2].

The utilization of geothermal energy according to the geological conditions are categorized as:

1. High-temperature (enthalpy) geothermal systems with temperature greater than 180°C. This system depends on recent mantle hot spot anomalies and volcanos at depths over 3.5 km. Also, high-temperature geothermal systems related to rocks at depths approximately below 3.5 km.

	(a)	(b)	(c)	(d)	(e)
Low enthalpy resources	<90°C	<125°C	<100°C	≤150°C	≤190°C
Middle enthalpy resources	90–150°C	125 – 225°C	100–200°C	—	—
High enthalpy resources	>150°C	>225°C	>200	>150°C	>190°C

a) Muffler and Cataldi (1978); b) Hochstein (1990); c) Benderitter and Cormy (1990); d) Nicholson (1993); e) Axelsson and Gunnlaugsson (2000).

Table 1.
Classification of geothermal resources based on temperature [3].

2. Middle-temperature (enthalpy) systems between 100 and 180°C.

3. Low-temperature (enthalpy) geothermal systems with temperature lesser than 100°C.

Geothermal systems with middle and low-temperature are formed by decaying of radioactive isotopes and they conclude aquifers which recharge by heated water circulation. **Table 1** tabulates the classification of the geothermal resources based on temperature [3].

Geothermal energy can be utilized for both purposes – direct heat and electricity generation. As of the end of 2019, geothermal energy was used in 88 countries around the world with an annual energy consumption of around 1,020,887 TJ or 283,580 GWh [4]. According to the International Energy Agency the electricity production from geothermal energy will be increased to 1400 TWh/y and the direct use to 1600 TWh/y by 2050 [5].

The main advantage of the geothermal energy is being clean as other types of renewable sources. Other advantages include: reliability, environment-friendliness, relatively low cost of generated energy, high usage factor for the geothermal power plant operation that distinguishes the geothermal energy from other renewable energy sources. However, the heat and electricity generated by the geothermal energy should be directly utilized locally and cannot be transported.

2. Geothermal power plants

In work [6], modeled types of geothermal resources are given, as shown in **Table 2**. In general, geothermal resources under 150°C are more suitable for direct use such as heating and cooling, whereas the resource of above 150°C is exploited for electricity generation. However, modern power conversion technologies allow to generate an electricity from low temperature resources up to 150°C [3].

A steam or hydrocarbon vapor are used to generate electricity from geothermal energy. While the vapor-dominated resource is applied directly, the hot-water dominated resource should be flashed by reducing the pressure to convert a steam [7]. **Table 3** provides the basic technologies and the common applications under different temperatures of geothermal fluid.

As mentioned above, the geothermal source in view of steam or hydrocarbon vapor can be used to generate electricity. In this respect, geothermal power plants (GPPs) operation is similar to those of steam power plants. However, unlike the conventional steam power plants, the geothermal power ones use natural steam of earth. History of the first GPP started from 1904 at Larderello, Italy,

Type of geothermal resource	Temperature
Hot-water dominated	20–350°C
Vapor-dominated	≈240
Sedimentary basin	20–150°C
Radiogenic	30–350°C
Geopressured	90–200°C
Solidified (hot dry rock)	90–650°C
Part still molten (magma)	>600 °C

Table 2.
Types of geothermal resources for energy utilization.

Reservoir temperature	Geothermal fluid	Application proposes	Technology
High temperature, >220°C	Water or steam	Direct-in use	Heat exchangers; Heat pumps
		Power generation	Flash steam: combined cycle (flash and binary)
Medium temperature, 100–220°C	Water	Direct-in use	Heat exchangers; Heat pumps
		Power generation	Binary cycle
Low temperature, 30–150°C	Water	Direct-in use	Heat exchangers; Heat pumps

Table 3.
Frequently used technologies for geothermal energy.

where generator was tested to produce electricity from geothermal source. Then, it was commercialized to power plant in 1911 [7].

The simplest operational conceptualization of the GPP is presented in **Figure 1**. In this case, a natural steam from the well is directly passed to a turbine, that drives a generator for the electricity production.

There are three basic types of GPPs:

1. Dry steam plants.

2. Flash steam plants.

3. Binary cycle power plants.

2.1 Dry steam plants

Dry steam plants are simple and more efficient type of GPP. This type of power plants was firstly deployed to generate electricity in Italy in 1911. However, dry steam plants are less available in the sense that the steam should be produced from vapor-dominated reservoirs which are of few numbers in the world.

To provide a high efficiency of turbine, the steam condensation is minimized during extension of fluid fraction in steam phase. In common, isentropic efficiency of modern dry steam plants is about 85%. According to a feasibility study to extract the maximum efficiency of these plants, the generation capacity should be at least 1 MWe [8].

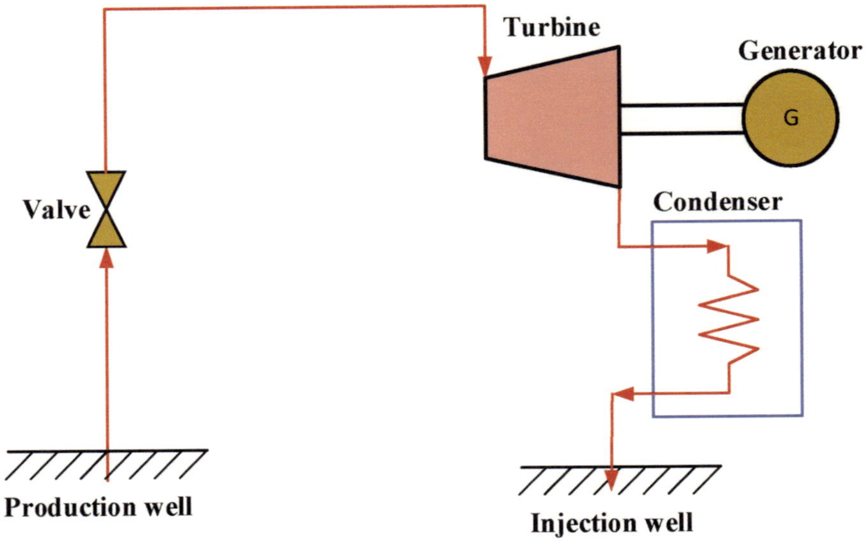

Figure 1.
Schematic diagram of dry steam power plant.

Based on [9], electricity can be generated from the dry steam in the following conditions:

1. The vapor-dominated source should be closer (approximately with 5 km depth) to the surface to raise the hot water to the boiling point.

2. There should be enough opening above the geothermal fluid source to allow the vapor to drop to the surface over a long period of time by reducing the level of liquid significantly.

The schematic diagram of dry steam power plant is illustrated in **Figure 1**. The dry steam plant operation is based on the following: the water and steam flow from geothermal production wells are transmitted through valve to turbine and spin a steam turbine by converting thermal and kinetic energy to electrical energy.

2.2 Flash steam plants

Currently, most of GPPs are flash steam plants. They use geothermal reservoirs with mixture sources, i.e., vapor and liquid-dominated (water) to generate an electricity. It means that the temperature and enthalpy of vapor and waters are lesser than the critical point. Therefore, the flash units are generally basic approach to change over the geothermal energy into power. Firstly, in this system the steam is separated from water using a cylindrical cyclonic pressure tank with a base loss of pressing factor. Then, the dry steam leaves the separator, flows to powerhouse and rotates the steam turbine. The vapor quality that defines the flashed fluid is given as:

$$x = \frac{m_{vapor}}{m_{liquid} + m_{vapor}} \tag{1}$$

where: m_{vapor} – mass of vapor and m_{liquid} – mass of liquid.

The vapor quality value changes from 0 to 1 and commonly given as percentage. When vapor quality between 0 and 1, wet steam is then obtained. When the vapor quality is equal to 1, it is called "the saturated vapor" state.

There are two types of flash steam geothermal plants:

- Single-flash steam GPP
- Double-flash steam GPP

In a single-flash steam GPP the mixture fluid is flashed only in one separator. Schematic process diagram of the single-flash power plant is shown in **Figure 2**. The process of electricity generation in this power plant is accomplished as follows: hot water from production well is piped to flash separator (FS) by decreasing its pressure. In FS the steam is separated from hot water and transmitted to the steam turbine to spin it and convert mechanical energy to electrical by a generator. While the cooled steam in a turbine condenses to the water by the condenser while a part of the liquid from FS are reinjected to injection well.

Unlike single-flash steam, in double-flash steam the flash process of the fluid is applied in two separators. Although these plants are more expensive and more labor-consuming, however, they are preferable than single-flash plants as they generate 15–25% more electricity for the same states of fluid reservoirs [8, 9].

Schematic operational diagram of the double-flash power plant is illustrated in **Figure 3**. The fluid flows from well to a high pressure flash separator where from a mixture fluid the steam is separated and is piped to a two-stage turbine; another part – saline liquid is throttled down to the second separator. In the low pressure separator like the first one a partly boiled liquid again is separated to a steam and water. As a result, the steam gets directed to the low-pressure turbine. By keeping the pressure in the condenser, the steam from the low-pressure turbine is cooled using a sprayed cold water. Then, water is reinjected to injection well, as well as the cold water from the condenser.

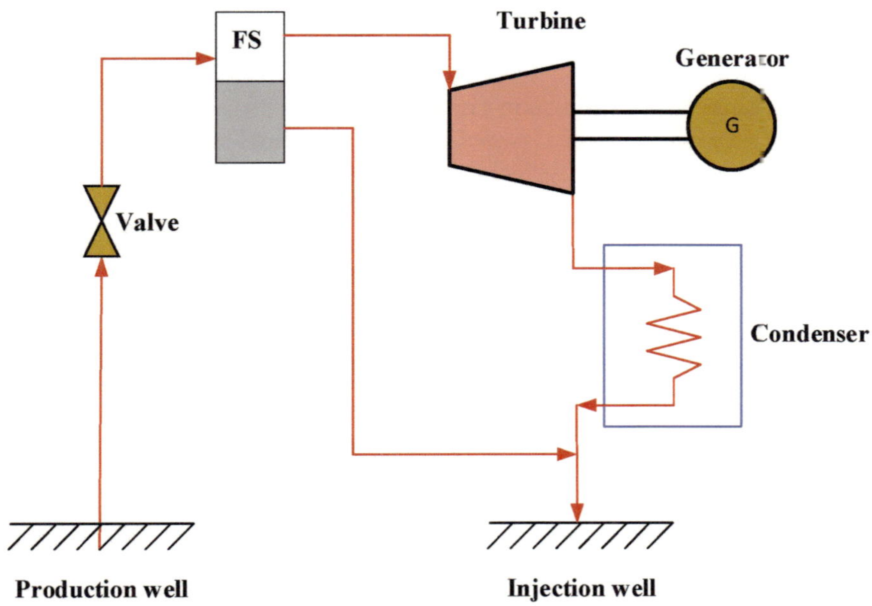

Figure 2.
Schematic diagram of single-flash power plant.

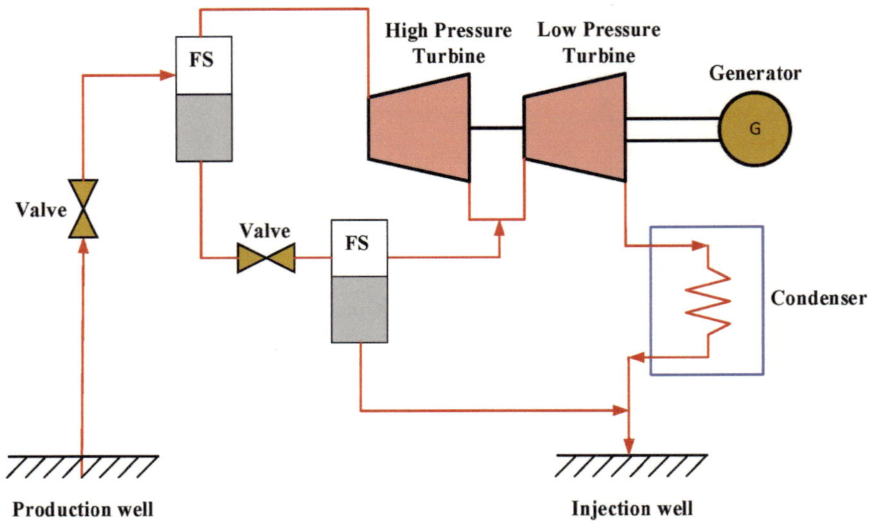

Figure 3.
Schematic diagram of double-flash power plant.

2.3 Binary cycle power plants

"Binary" cycle refers to as a secondary separate cycle. For the geothermal resource, binary indicates that the geothermal fluid (water/steam) never comes in a contact with the prime mover. Geothermal binary power systems are suitable for electricity production from low underground heat source [10]. The binary plant in Alaska, as an example, utilizes a geothermal resource of 57°C [11]; yet generally, binary system designs can exploit an inlet temperature range between 80 and 170°C [3]. The secondary fluid, known commonly as working fluid, in the binary geothermal system operates in a conventional Ranking cycle; and the binary cycle is known as an organic Ranking cycle (ORC) when the used working fluid is organic [12]. In binary ORC power plants the geothermal fluid passes through a heat exchanger to heat another working fluid of a low boiling point e.g., pentane, zeotropic mixtures, etc. which in turn vaporizes and drives a turbine [13]. Electrical production through a closed-loop binary unit is shown in **Figure 4**.

The standard working mechanism of a basic geothermal binary system can be summarized as: when the geothermal brine is pumped through the production well,

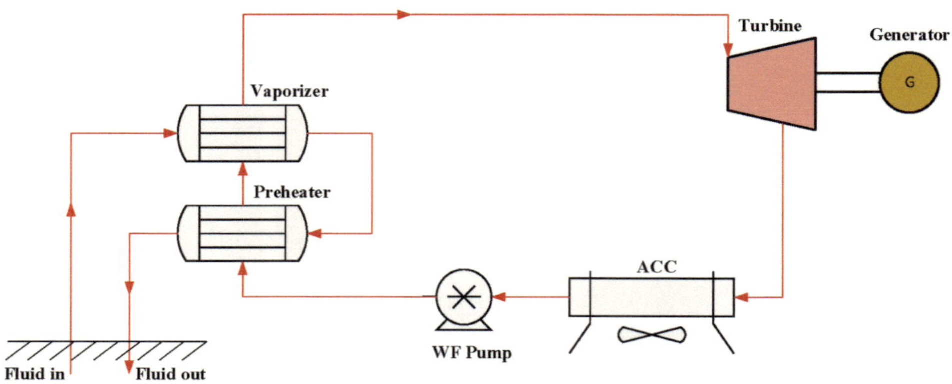

Figure 4.
Illustrative schema for a binary geothermal plant.

the heat extraction process is accomplished after passing through different components of the primary cycle. The geothermal fluid is initially filtered via sand removers to pass through the heat exchanger i.e., the evaporator/vaporizer and preheater, and finally pumped back into the reservoir by the injection well. On the other side of the secondary cycle, the pressurized working fluid turns into boiling state in the preheater. It then exits the vaporizer as a saturated vapor that subsequently expands in the turbine driving a power generator. The low-pressure working fluid vapor exiting the turbine is finally condensed in the ACC (air-cooled condenser) and pumped back to the vaporizer, closing the loop system and repeating the process continuously. Thus, the thermodynamic process of the low-boiling-point working fluid starts when it expands into the turbine in saturation vapor state, and completed when it is cooled through the condenser and pumped back (as a saturated liquid fluid) to the heat exchanger to emerge as a saturated vapor again [14, 15].

The efficiency of geothermal ORC, for high enthalpy field, can go as high as 23% [16]. However, cycle configuration plays a key role in thermodynamics of a binary power plant. Many performance and optimization studies have been recently carried out to examine the optimal configuration of ORC geothermal power facilities [17, 18] as well as on the investigation of optimal working fluids in ORCs [19–21]. In Ref. [22], the researchers investigated the performance of three configurations of ORC for binary geothermal power plants; simple ORC, regenerative ORC and ORC with Internal Heat Exchanger (ORC-IHE). It is concluded that the ORC-IHE outperforms the other configurations from the thermodynamic perspective while the simple ORC had the highest value of net output power. The 2-stage designs of a binary cycle yield higher net electrical power output and thermal and exergy efficiencies than the 1-stage counterparts [23].

As the thermal energy extracted from underground field is conveyed to a second working fluid; therefore, selection of such working medium plays an important role on the system design, performance, and economics. The optimal choice of the working fluid for a binary cycle must consider the thermodynamic characteristics of both geofluid and working fluid, safety of use, health and environmental impact [9]. Various objective functions have been used in literature for working fluid selection, such as the net power output [24], ratio of net power output to heat exchanger area [25], first or second law efficiencies [26] and volumetric expanders [27]. In [28] the authors conducted a comparative study of several working fluids, such as water, coolants and some hydrocarbons, for a Rankine cycle operating at low temperature. The study concluded that using organic working fluids, the Rankine cycle achieved good efficiencies for the recovery of low enthalpy resources. An optimization study conducted by [29] revealed that the n-pentane working fluid produced the highest first and second law efficiencies for a binary ORC plant. The study in [29] explored the thermodynamic performance of 20 working fluids for a binary ORC and found out that R123, R141b and ethanol are the most appropriate for small scale domestic Combined Heat and Power (CHP) applications. CHP or cogeneration plants are efficient technology that produces both electricity and thermal energy at considerably higher efficiency than its counterpart of only-electricity or only-heat systems.

It was reported that the operational parameters of a binary plant (such as air mass flow rate, mass flow rate of organic medium and inlet turbine pressure) and plant performance i.e., net power output degrade over the plant lifetime [14]. In order to maintain the plant performance over its life span, the mass flow rates of organic fluid and air cooling should be adjusted. In addition, the plant design can be modified by placing a recuperator and reducing the heat transfer area of vaporizer and preheater.

Besides the standard binary geothermal power system, advanced configurations of geothermal energy conversion systems have been also well investigated. This includes: hybrid single-flash and double-flash systems, hybrid flash-binary configuration and hybrid fossil geothermal technology [23, 30]. In addition, the development of hybrid power systems integrating geothermal plants with biomass, fuel cells, wind, solar systems and waste-to-energy (WTE) technologies has been gaining a lot of interest [31–35].

When compared to single-flash and double-flash cycles, the binary ORC plant attained the highest thermal efficiency and output power among the three geothermal power plants [36]. An exergoeconomic investigation between double-flash and single-flash/ORC combined cycles revealed that the single-flash/ORC integrated cycle offers the highest energy and exergy efficiencies [37]. Based on a comparison between various types of geothermal power plants in terms of energy and exergy metrics, it was concluded that the combined flash-binary cycle with R123 working fluid, at a temperature 230°C and mass flow rate of geothermal heat source 1 kg/s, has the highest amount between various investigated configurations with a maximum thermal efficiency of 11.81% [38].

A combined flash-binary ORC power unit is schematically shown in **Figure 5**. The working mechanism of the plant is as follows; firstly, the geofluid is throttled in a valve to a lower pressure (point 2), then the obtained two-phase fluid is decomposed into saturated liquid and saturated steam by getting through the flash chamber i.e., separator (3 and 4). The extracted saturated steam drives a prime mover that is connected to a power generator (5); the steam turbine exhaust is then cooled in the condenser (6). On the other hand, the saturated liquid of the separator enters the heat exchanger to give off heat to the binary unit (7). The exit mixed stream of condenser and heat exchanger is then injected back into the ground (8). The pump

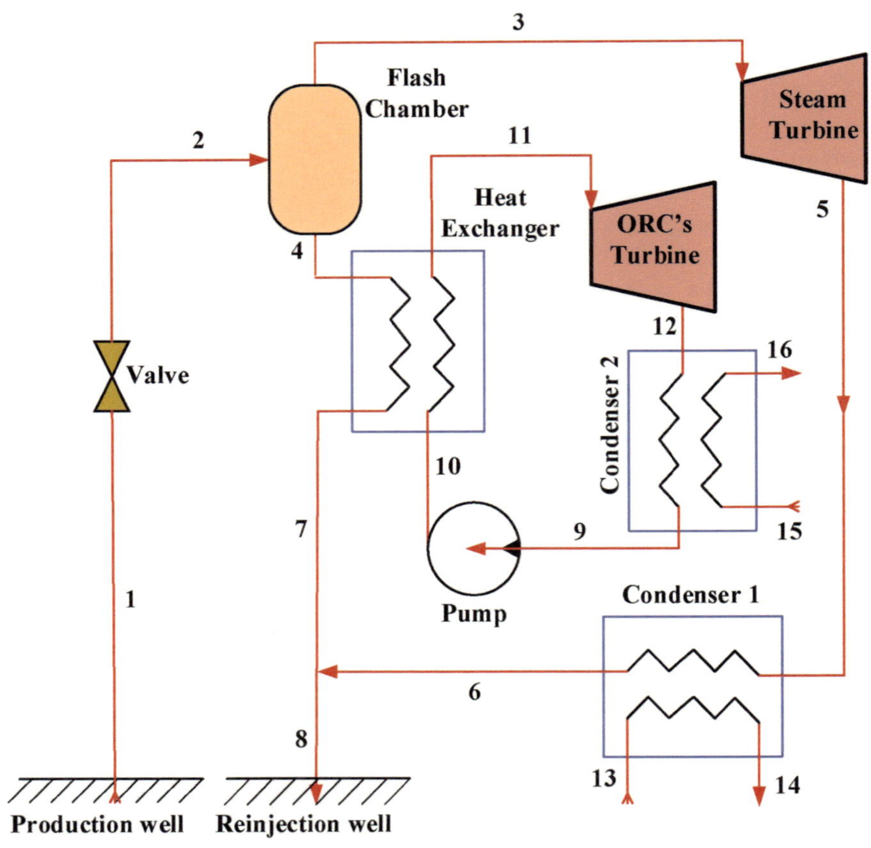

Figure 5.
Schematic diagram of the single flash/ORC combined cycle.

within the binary ORC pressurizes the organic working fluid to a high level (10), that is in turn be in a form of saturated vapor when thermally exchanged with the heat of saturated geofluid stream in the evaporator (11). The saturated vapor is expanded into the steam turbine and delivers work to produce further electricity (12). Finally, water flow in the condenser condenses the superheat vapor and exits as a saturated liquid (9) [39].

As mentioned earlier, integration of multiple generators of different technologies especially of renewable ones has been attracting a considerable attention. The synergy offers cost competitiveness, greater overall efficiency and a higher capacity factor as compared to a single source power supply [40]. For instance, hybridization of geothermal with concentrating solar power (CSP) can overcome several challenges encountered by standalone geothermal plants [41]. The concept is that as the ambient temperature increases with the progress of the day the hourly output of a standalone geothermal plant decreases. Nevertheless, CSP involvement can handle this issue as its output increases with a rise in the ambient temperature and more than 70% in annual energy output could be attained [42]. Generally, CSP can be incorporated in the geothermal plant both in the preheating or the superheating configuration [43].

3. Worldwide status of geothermal power production

By the end of 2018, the cumulative global installed capacity of geothermal power amounted to 13.28 GW generated an annual electrical energy of about 86 TWh. Geothermal resource has contributed significantly in electricity production in some regions; 17% of New Zealand's electricity production and 31% of Iceland's electricity production was met by geothermal in 2018 [44]. In 2020, the global additions of geothermal capacity are estimated at 300 MW. Indonesia and Turkey led the new development, with 145 MW and 70 MW of capacity added for the two countries, respectively. The technology is forecast to reach 16.5 GW of aggregate capacity worldwide by 2022 [45]. While United Stated of America has been at the top world rank in terms of geothermal power capacity [46], Turkey, Kenya, Indonesia and the Philippines would be responsible for most of the technology growth and continue to lead capacity additions beyond 2022. Geothermal power cumulative capacity and additions in leading countries are depicted in **Figure 6** [44].

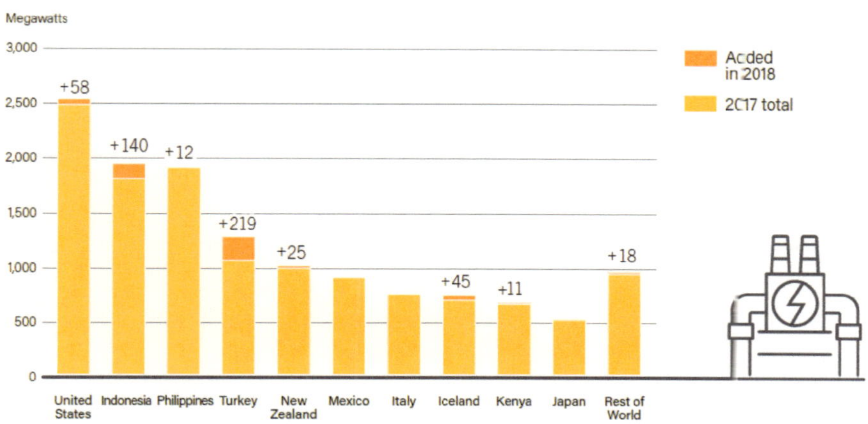

Figure 6.
Geothermal power capacity and additions in top countries by 2018 [44].

Geothermal technology exploitation depends primarily on the resource potential and economic considerations, but generally, most existing geothermal facilities worldwide use flash or dry-steam technologies. However, on global scale, binary-cycle technology has been the fastest progressing technology in recent time, due in part to rising utilization of relatively low-temperature resources [44]. Direct Use i.e., thermal energy consumption of geothermal energy technology is one of the common and versatile form of utilizing the underground heat [47]. The installed global total of geothermal power capacity for Direct Use sectors is estimated at 26 GWt at the end of 2018 [44].

The distribution of geothermal Direct Use applications is categorized as: 58.8% for heat pumps, 18% for bathing and swimming, 16% for space heating, 3.5% for greenhouse heating, 1.6% for industrial applications, 1.3% for raceway heating and aquaculture pond, while the remainder goes to other applications such as agricultural drying and snow melting and cooling, etc. [47]. Geothermal heat pumps have the greatest geothermal utilization worldwide, accounting for 59.2% of the annual energy use and 71.6% of the installed capacity by 2020. The installed capacity for geothermal based heat pumps approaching 77,547 MWt mostly built in North America, Europe and China. The size of individual systems ranges from 5.5 kW for residential installation to greater than 150 kW for commercial and institutional units [47]. Most heat pumps systems in Europe are sized for the heating load that are designed to meet the base load, with surging by fossil fuels. In Finland, as an example, some of these units reached an operation of up to 3,000 equivalent full-load heating hours per year i.e., a capacity factor of 0.34.

Space heating, including district heating and individual space heating, has now an installed capacity of 12,768 MWt and an annual energy use of 162,979 TJ/yr. In terms of annual energy use, the leaders are Turkey, Russia, Japan, the United States,

Utilization	Installed Capacity (MWt)	Capacity Factor	Leaders in annual energy utilization (TJ/yr)
Geothermal heat pumps	77,547	0.245	United States, China, Germany, Sweden and Finland
Space heating	12,768	0.405	Turkey, Russia, Switzerland, the United States and Japan
Greenhouse heating	2,459	0.462	China, Turkey, Netherlands, Hungary and Russia
Aquacultural pond heating	950	0.463	United States, China, Italy, Iceland and Israel
Agricultural drying	257	0.435	China, Hungary, France, Japan and United States
Industrial uses	852	0.610	New Zealand, China, Russia, Iceland and Hungary
Bathing and swimming	12,253	0.473	Japan, China, Turkey, Mexico and Brazil
Snow melting	435	0.189	Iceland, Japan, United States, Argentina and Slovenia
Other	106	0.584	New Zealand (irrigation and frost protection), Japan (cooking) and Kenya (boiling water)
Total	**107,727**		

Table 4.
MWt and leading states for various categories of direct-utilization for the year 2020 [4].

and Switzerland, accounting for about 75% of the world's individual space heating and nearly 90% of the world's total use in district heating. Deep Direct Use or what is alternatively called Cascaded Use offers large-scale viable systems that optimize the value stream of lower temperature resources through a multiple of purposes, from electricity production to direct cooling and heating, commercial and industrial applications, etc. [46]. **Table 4** provides a summary of the installed capacity factor (in MWt), and leading states for various categories of Direct Use for the year 2020.

4. Economic and environmental impacts of geothermal energy

In addition of being harnessed for the production of electrical energy, geothermal energy can also be utilized for various thermal applications including industrial heat input and space heating. Geothermal waters are highly beneficial for health and well-being, treat arthritis and skin diseases. Production of freshwater and minerals exploiting the hot reinjected brine of geothermal resource is also a viable and economic option [4]. Furthermore, geothermal energy resources improve the security and defense of the country through their exploitation in military facilities such as heating of runways and heliports and wide range applications of heat pumps [48]. For the above-mentioned advantages, many countries offer incentives for the use of the technology. For instance, Switzerland raised its geothermal power Feed-in Tariff (FiT) from USD 0.48 to USD 0.54 per kWh [44].

Nevertheless, the main barriers for the development of geothermal industry include high resource and exploration risk, overall high development cost particularly drilling, economic risk associated with long project lead-times and inadequate financing and grant support, as well as lack of clear policy and regulatory frameworks. Indonesia, for example, did not meet its 2018 targets for investment acceleration in geothermal due largely to drilling delays by developers. The economic recession stemming from COVID-19 crisis has enormously affected the technology progress even for the pioneered geothermal energy users such as Italy, the United States and New Zealand that have not witnessed significant growth in recent years [44]. The global crisis has caused deferrals of strategic decisions such as financing and disruptions to the global supply chain for materials and machinery. On the other hand, it is reported that the availability of better data about geothermal resources facilitates attracting new investors and developing new projects [49].

The per MW cost of geothermal power unit hits USD 7 million [44]. Global range of electricity cost and its weighted average of various technologies are tabulated in **Table 5**. As can be seen, renewable technologies are competing with fossil fuels, while geothermal projects are not far behind (at USD 72/MWh). Because of its adequacy for relatively low-temperature resources and applicability for both power and heat, among the key players of geothermal binary technology are Exergy (Italy), Ormat Technologies (United States), and Turboden (Italy, a subsidiary of Mitsubishi Heavy Industries of Japan) [44].

An economic analysis performed by [23] indicated that the economic performance of the Rankine cycle depends greatly upon the type of working fluid and cycle configuration. The results also showed that a standard Rankine cycle with a 2-stage turbine using n-pentane is the most thermo-economic design for the particular brine resource and re-injection conditions. For binary ORC plants, the simple ORC offers the lowest total capital investment and the shortest payback period as compared to the regenerative and the one with internal heat exchanger [22]. An economic assessment of double-flash geothermal power cycle and single-flash-ORC combined cycles showed that the former configuration attains the minimum unit cost of produced power [37]. In another economic analysis

Technology	COE 5th and 95th Percentiles (USD/kWh)	Global Weighted-Average COE(USD/kWh)	Change in COE 2017–2018
Geothermal	0.060–0.143	0.072	−1%
Hydropower	0.030–0.136	0.047	−11%
Onshore wind	0.044–0.100	0.056	−13%
Offshore wind	0.102–0.198	0.127	−1%
Solar Photovoltaic	0.058–0.219	0.085	−13%
Concentrating Solar Power	0.109–0.272	0.185	−26%
Bioelectricity	0.048–0.243	0.062	−14%
Fossil fuels	0.049–0.174	0.049–0.174	N/A

Table 5.
Global electricity cost of different technologies as of 2018 [44, 50].

of a hybrid CSP-binary geothermal power plant showed that the levelized cost of electricity can be reduced by 2% for the hybrid system in comparison to the stand-alone geothermal system [51].

Land requirement of geothermal power plant is relatively much lower than other technologies. For example, a single-flash geothermal unit needs approximately 1200 m^2 per MW installation as compared to coal-fired facility which requires 40,000 m^2/MW and a photovoltaic plant requirements of 66,000 m^2/MW [9]. On the contrary, this economic advantage in land aspect is challenged with other environmental concerns such as water usage and its pollution, visual and noise pollution, greenhouse gas emissions, and loss of natural beauty. However, methods to alleviate these environmental concerns include reinjection for surface water pollution, the use of silencers for noise pollution and air-cooled condensers for water usage [52].

The great benefits of increased implementation of geothermal technology include the high reliability and the feasible functionality over 7000 h per year, which is a crucial issue for electrical utility grids. If the reservoir is appropriately managed i.e., the reservoir water balance, then the sustainability of geothermal power plants is guaranteed [53]. However, the release of NCGs to the environment has become a critical factor for a geothermal power plant. NCGs are naturally found in geothermal reservoirs and can contain several types of pollutants, dominantly carbon dioxide (CO_2), in addition to ammonia (NH_3), hydrogen sulfide (H_2S) and heavy metals. Although CO_2 production of the geothermal power plants is much lower than fossil fuel generation units, they still emit averagely 400 g CO_2/kWh and may be higher depending upon the chemical composition of the reservoir and the conversion technology [54].

Treatment of NCGs has been a hot topic in the industry world and research community. After segregation of the NCGs, geothermal fluids are reinjected into the reservoir. This is a demonstrated design in geothermal power plants. The Sinem ORC geothermal power plant rated at 24 MW and located in Aydın province, Turkey re-injects about 70% of the heat drawn fluid, after being condensed at 70°C, into re-injection well whereas the remaining 30% is emitted to atmosphere as NCGs [55]. However, an emerging technology based on reinjection of NCGs is still under development intending to minimize the environmental footprint, handling the emissions of H_2S and CO_2 [53]. It is confirmed that complete reinjection of NCGs using binary ORC geothermal power plant assures the sustainability of geothermal resource [56]; in this regard, some advancements have been achieved in Iceland geothermal plants [57]. More funds are dedicated for eliminating the environmental effect of geothermal industry. The European Commission, for example, awarded the Geothermal Emission Control (GECO) project, USD 18.3 million to advance research on reinjection of harmful gases such as CO_2 and H_2S from open-loop geothermal plants [44].

Feasibility of hydrogen production by means of geothermal resource has been also explored. In [58] the authors conducted a thermo-economic cost assessment of electrolysis based hydrogen production powered via a binary geothermal unit. The analysis argues that for a geothermal heat of 160°C and a flow rate of 100 kg/s, hydrogen can be obtained at a level of 0.253 g per kilogram of geothermal water. It was also found that unit exergetic costs of electricity and hydrogen are 0.0234 \$/kWh 2.366 \$/kg H_2, respectively. Koroneos et al., [59] demonstrated the technical feasibility (based on efficiencies and exergy indicators) of installing a 2.1 MW binary geothermal power plant at in Nisyros Island, Greece. The proposed facility could reach up to 10 MW of total installed capacity in the future and capable of supplying a substantial amount of electricity thereby reducing the reliance of the island to the diesel power generation thus the gain from an environmental point of view is guaranteed.

Energy savings from utilizing Direct Use geothermal energy amounts to 81 million tonnes (596 million barrels) of equivalent oil yearly. This eventually prevented 78.1 million tonnes of carbon and 252.6 million tonnes of CO_2 from being released to the atmosphere [4]. Geothermal technology based freshwater production is one of the most economic renewable and clean production alternative [60].

Hybrid geothermal-fossil fuel power system, for low-enthalpy geothermal resources, is a practical alternative to reduce the extensive use of fossil fuels and associated emissions. The research study in [61] analyzed a 500 MW combined geothermal-coal power plant with a 210°C geothermal temperature and 400 kg/s brine flow rate. The study claimed that up to 0.3 million tonnes of coal can be saved per year in addition to annual reductions of up to 0.72 million tonnes of greenhouse gas emissions. Economically, a drop of 33–87% in energy cost is reachable in comparison to a sole geothermal power unit. A study conducted by [51] revealed that the hybridization of binary ORC with CSP system decreases the levelized energy cost by 2% and when the ORC geothermal configuration is optimized an 8% drop in the levelized energy cost is achieved.

In [55], an exergoenvironmental analysis is performed from the perspective of environmental impact. The study came to a conclusion that 98% of total environmental impact for the geothermal power system is caused by exergy destruction of the equipment involved; and as a treatment, exergetic efficiency for equipment should be improved rather than construction, operation/maintenance and disposal changes of facility equipment. The study advices obtaining a higher capacity plant by having a better condenser performance and enhancing the efficiency of the vaporizers and the pumps.

5. Conclusion

The population's growth and economic development in many countries require an increase in demand for electrical energy. On the other hand, meeting the need for electric energy society in the future should consider the limitations of non-renewable resources while providing energy sustainability and significantly reducing the negative impact on the environment. In this case, the role of renewable energy resources should thus be a priority.

The geothermal energy is one of the sustainable and environmentally friendly sources. Also, it is of a high usage factor and reliability. While geothermal dry steam and flash power plants utilize medium to high enthalpy underground source to generate electricity, geothermal binary power systems are suitable for electricity production from low underground heat source; generally, an inlet temperature range between 80 and 170°C. The secondary working fluid in the binary geothermal system operates in a conventional Ranking cycle. The main feature of the working fluid is having a low boiling point. The configurations of the cycle and the type of working fluid should be optimized for the sake of obtaining the best possible thermodynamic performance and efficiency of the geothermal binary facility. The binary Rankine cycle that used an organic working fluid has demonstrated its efficiency and practicality. Among the reported configurations of ORC for binary geothermal power plants include simple ORC, regenerative ORC and ORC-IHE.

Besides the standard binary geothermal power system, advanced configurations of geothermal energy conversion systems have been also well investigated. They include hybrid single-flash and double-flash systems, hybrid flash-binary configuration and hybrid fossil geothermal technology. In addition, the development of hybrid power systems integrating geothermal plants with biomass, fuel cells, wind, solar systems, and WTE technologies has been gaining a lot of interest. It is reported that the binary ORC plant, when compared to single-flash and double-flash cycles, attains the highest thermal efficiency and output power. On the other hand, the single-flash/ORC integrated cycle offers the highest energy and exergy efficiencies as compared to double-flash counterpart.

The accumulated global installed capacity of geothermal power amounted to 13.28 GW by the end of 2018 and is forecast to reach 16.5 GW worldwide by 2022. United Stated of America has been standing at the top world rank in terms of geothermal power capacity. On the other hand, Turkey, Kenya, Indonesia and the Philippines would be responsible for most of the technology growth and continue to lead capacity additions beyond 2022. Although most existing geothermal facilities worldwide use flash or dry-steam technologies, binary-cycle technology has been the fastest progressing technology in recent time. Direct Use of geothermal energy technology is one of the common forms of utilizing the underground heat. The distribution of geothermal Direct Use applications, relative to their widespread use, are: heat pumps, bathing and swimming, space heating, greenhouse heating, industrial applications, raceway heating and aquaculture pond, agricultural drying and snow melting and cooling, etc.

The main barriers for the development of geothermal industry include high resource and exploration risk, overall high development cost particularly drilling, economic risk associated with long project lead-times and inadequate financing and grant support. In addition, the global crisis COVID-19 has caused deferrals of strategic decisions such as financing and disruptions to the global supply chain for materials and machinery. The economic feasibility of the geothermal power plants depends greatly upon the type of working fluid and cycle configuration. Geothermal power plants have relatively much lower land requirement than other technologies. Methods to alleviate the environmental concerns of geothermal power systems include reinjection for surface water pollution, the use of silencers for noise pollution and air-cooled condensers for water usage. NCGs accompanying the geothermal power generation can be alleviated by reinjecting geothermal fluids into the reservoir.

As a future work, an optimized hybridization of geothermal energy with other renewables needs further exploration and demonstration, either for small scale CHP or cascaded applications. As the availability of better data about geothermal resources facilitates the attraction of new investors and developing new projects, more deeper studies and financial support is critical to assess sites of appropriate geothermal potential. This is of significance to make the cost of geothermal electricity more competitive to conventional and other cheaper renewable electricity. Advanced technologies to eliminate the environmental effect of geothermal power plants are also required including complete reinjection of NCGs that needs further research, fund, and practical demonstration.

Author details

Ziyodulla Yusupov[1]* and Mohamed Almaktar[2]

1 Karabuk University, Karabuk, Turkey

2 College of Electrical and Electronics Technology, Benghazi, Libya

*Address all correspondence to: ziyodulla@gmail.com

References

[1] Fridleifsson IB, Bertani R, Huenges E, Lund JW, Ragnarsson A, Rybach L. The possible role and contribution of geothermal energy to the mitigation of climate change. In: IPCC Scoping Meeting on Renewable Energy Sources, Proceedings, 20-25 January 2008; Luebeck, Germany: 2008. p. 59-80

[2] Pollack HN, Hurter SJ, Johnson; Johnson JR. 1993; Heat flow from the Earth's interior: Analysis of the global data set, Rev. Geophys., 30 (3), pp. 267-280, DOI:10.1029/93RG01249

[3] Dickson M, Fanelli M, Geothermal energy: utilization and technology. Routledge; 2013

[4] Lund W, Toth AN. Direct utilization of geothermal energy 2020 worldwide review. Geothermics. 2021; 90: 1-39. DOI: 10.1016/j.geothermics.101915

[5] Technology roadmap. Geothermal heat and power. Technical report, 2011. International Energy Agency. Paris, France

[6] White DE, Williams DL. Assessment of geothermal resources of the United States – 1975. U.S. Geological Survey Circular 727, U.S. Government Printing Office. 1975. 155 p

[7] Power Stations Using Locally Available Energy Sources: A Volume in the Encyclopedia of Sustainability Science and Technology Series, Ed. Bronicki LY, Springer, New York, 2018

[8] Mondejar ME, Chamorro CR. Geothermal Power Technologies. In: M. A. Abraham (Ed.), Encyclopedia of Sustainable Technologies. Elsevier. 2017. p. 51-61.

[9] DiPippo R. Geothermal Power Plants: Principles, Applications, Case Studies and Environmental Impact: Fourth Edition. 2016. 762 p.

[10] Almaktar M, Shaaban M. Prospects of renewable energy as a non-rivalry energy alternative in Libya. Renewable and Sustainable Energy Reviews. 2021;143:110852. DOI: 10.1016/j.rser.2021.110852

[11] Erkan K, Holdmann G, Benoit W, Blackwell D. Understanding the Chena Hot Springs, Alaska, geothermal system using temperature and pressure data from exploration boreholes. Geothermics. 2008; 37(6):565-585. DOI: 10.1016/j.geothermics.2008.09.001

[12] Zhao Y, Wang J. Exergoeconomic analysis and optimization of a flash-binary geothermal power system. Appl. Energy. 2016;179:159-170. DOI: 10.1016/j.apenergy.2016.06.108

[13] Zarrouk SJ, Moon H. Efficiency of geothermal power plants: A worldwide review. Geothermics. 2014;51:142-153. DOI: 10.1016/j.geothermics.2013.11.001

[14] Budisulistyo D, Wong CS, Krumdieck S. Lifetime design strategy for binary geothermal plants considering degradation of geothermal resource productivity. Energy Convers. Manag. 2017;132:1-13. DOI: 10.1016/j.enconman.2016.10.027

[15] Moya D, Aldás C, Kaparaju P. Geothermal energy: Power plant technology and direct heat applications. Renew. Sustain. Energy Rev. 2018; 94:889-901. DOI: 10.1016/j.rser.2018.06.047

[16] Bonafin J, Pietra C, Bonzanini A, Bombarda P. CO2 emissions from geothermal power plants: evaluation of technical solutions for CO2 reinjection. In: Proceedings of European Geothermal Congress (EGC'19); 11-14 June 2019; Den Haag, The Netherlands: 2019. p. 1-9

[17] Erdeweghe SV, Bael JV, Laenen B, D'haeseleer W. Optimal configuration

for a low-temperature geothermal CHP plant based on thermoeconomic optimization. Energy. 2019;179:323-335. DOI: 10.1016/j.energy.2019.04.205

[18] Erdeweghe SV, Bael JV, Laenen B, D'haeseleer W. Optimal combined heat-and-power plant for a low-temperature geothermal source. Energy. 2018;150:396-409. DOI: 10.1016/j.energy.2018.01.136

[19] Saleh B, Koglbauer G, Wendland M, Fischer J. Working fluids for low-temperature organic Rankine cycles. Energy. 2007;32(7):1210-1221. DOI: 10.1016/j.energy.2006.07.001

[20] Heberle F, Brüggemann D. Thermo-economic evaluation of organic rankine cycles for geothermal power generation using zeotropic mixtures. Energies. 2015;8(3):2097-2124. DOI: 10.3390/en8032097

[21] Uusitalo A, Honkatukia J, Turunen-Saaresti T, Grönman A. Thermodynamic evaluation on the effect of working fluid type and fluids critical properties on design and performance of Organic Rankine Cycles. J. Clean. Prod., 2018;188:253-263. DOI: 10.1016/j.jclepro.2018.03.22

[22] Zare V. A comparative exergoeconomic analysis of different ORC configurations for binary geothermal power plants. Energy Convers. Manag. 2015;105:127-138. DOI: 10.1016/j.enconman.2015.07.073

[23] Budisulistyo D, Krumdieck S. Thermodynamic and economic analysis for the pre-feasibility study of a binary geothermal power plant. Energy Convers. Manag. 2015;103:639-649. DOI: 10.1016/j.enconman.2015.06.069

[24] He C, et al. The optimal evaporation temperature and working fluids for subcritical organic Rankine cycle. Energy. 2012;38(1):136-143. DOI: 10.1016/j.energy.2011.12.022

[25] Hettiarachchi HDM, Golubovic M, Worek WM, Ikegami Y. Optimum design criteria for an Organic Rankine cycle using low-temperature geothermal heat sources. Energy. 2007;32(9):1698-1706. DOI: 10.1016/j.energy.2007.01.005

[26] Li J, Pei G, Li Y, Wang D, Ji J. Energetic and exergetic investigation of an organic Rankine cycle at different heat source temperatures. Energy. 2012;38(1):85-95. DOI: 10.1016/j.energy.2011.12.032

[27] Kolasinski P. The method of the working fluid selection for organic Rankine cycle (ORC) systems employing volumetric expanders. Energies. 2020;13(3):573. DOI: 10.3390/en13030573

[28] Vélez F, Segovia JJ, Martín C, Antolín G, Chejne F, Quijano A. Comparative study of working fluids for a Rankine cycle operating at low temperature. Fuel Process Technol. 2012;103:71-77. DOI: 10.1016/j.fuproc.2011.09.017

[29] Mikielewicz D, Mikielewicz J. A thermodynamic criterion for selection of working fluid for subcritical and supercritical domestic micro CHP. Appl. Therm. Eng. 2010;30(16):2357-2362. DOI: 10.1016/j.applthermaleng.2010.05.035

[30] Liu Q, Shang L, Duan Y. Performance analyses of a hybrid geothermal-fossil power generation system using low-enthalpy geothermal resources. Appl. Energy. 2016;162:149-162. DOI: 10.1016/j.apenergy.2015.10.078

[31] Briola S, Gabbrielli R, Bischi A. Off-design performance analysis of a novel hybrid binary geothermal-biomass power plant in extreme environmental conditions. Energy Convers. Manag. 2019;195:210-225. DOI: 10.1016/j.enconman.2019.05.008

[32] Li K, Liu C, Jiang S, Chen Y. Review on hybrid geothermal and solar power

systems. J. Clean. Prod. 2020;250:119481. DOI: 10.1016/j.jclepro.2019.119481

[33] Kazmi SWS, Sheikh MI. Hybrid geothermal–PV–wind system for a village in Pakistan. SN Appl. Sci. 2019;1(7):1-15. DOI: 10.1007/s42452-019-0643-9

[34] Rahmanifard H, Plaksina T. Hybrid compressed air energy storage, wind and geothermal energy systems in Alberta: Feasibility simulation and economic assessment. Renew. Energy. 2019;143:453-470. DOI: 10.1016/j.renene.2019.05.001

[35] Olabi AG, Mahmoud M, Soudan B, Wilberforce T, Ramadan M. Geothermal based hybrid energy systems, toward eco-friendly energy approaches. Renew. Energy. 2020;147:2003-2012. DOI: 10.1016/j.renene.2019.09.140

[36] Jalilinasrabady S,Itoi R. Flash cycle and binary geothermal power plant optimization. Geothermal Resources Council Transactions. 2012;36:1079-1084

[37] Shokati N, Ranjbar F, Yari M. Comparative and parametric study of double flash and single flash/ORC combined cycles based on exergo-economic criteria. Appl. Therm. Eng. 2015;91(5):479-495. DOI: 10.1016/j.applthermaleng.2015.08.031

[38] Yari M. Exergetic analysis of various types of geothermal power plants. Renew. Energy. 2010;35(1):112-121. DOI: 10.1016/j.renene.2009.07.023

[39] Kolahi MR, Nemati A, Yari M. Performance optimization and improvement of a flash-binary geothermal power plant using zeotropic mixtures with PSO algorithm. Geothermics. 2018;74:45-56. DOI: 10.1016/j.geothermics.2018.02.004

[40] Qin J, Hu E, Li X. Solar aided power generation: A review. Energy Built Environ. 2019;1(1):11-26. DOI: 10.1016/j.enbenv.2019.09.003

[41] Cardemil JM, Cortés F, Díaz A, Escobar R. Thermodynamic evaluation of solar-geothermal hybrid power plants in northern Chile. Energy Convers. Manag. 2016;123(1):348-361. DOI: 10.1016/j.enconman.2016.06.032

[42] Zhou C, Doroodchi E, Moghtaderi B. An in-depth assessment of hybrid solar-geothermal power generation. Energy Convers. Manag. 2013;74:88-101. DOI: 10.1016/j.enconman.2013.05.014

[43] Wendt D, Mines G, Turchi C, Zhu G. Geothermal Risk Reduction via Geothermal/Solar Hybrid Power Plants: Final Report, Idaho National Laboratory. 2015. p. 156

[44] REN21.2019. Renewables 2019 Global Status Report (Paris: REN21 Secretariat). p. 336

[45] International Renewable Energy Agency. [Internet]. Available: www.irena.org. [Accessed: 12-Jan-2020]

[46] US Department of Energy." [Internet]. Available: https://www.energy.gov/eere/geothermal/electricity-generation. [Accessed: 23-Feb-2021]

[47] Capocelli M, Moliterni E, Piemonte V, de Falco M. Reuse of waste geothermal brine: Process, thermodynamic and economic analysis. Water (Switzerland). 2020;12(2):1-18. DOI: 10.3390/w12020316

[48] Sowiżdżał A, Górecki W, Chmielowska A. Geothermal energy as an opportunity to improve the security and defense of the country. In: Proceedings of European Geothermal Congress (EGC'19); 11-14 June 2019; Den Haag, The Netherlands: 2019. p. 1-8

[49] International Energy Agency [Internet]. Available: https://www.iea.org/. [Accessed: 13-Feb-2021]

[50] IRENA, "Renewable Power Generation Costs in 2018," 2019

[51] Ayub M, Mitsos A, Ghasemi H. Thermo-economic analysis of a hybrid solar-binary geothermal powerplant. Energy. 2015;87:326-335. DOI: 10.1016/j.energy.2015.04.106

[52] Bayer P, Rybach L, Blum P, Brauchler R. Review on life cycle environmental effects of geothermal power generation. Renewable and Sustainable Energy Reviews. 2013; 26:446-463. DOI: 10.1016/j.rser.2013.05.039

[53] Niknam PH, Talluri L, Fiaschi D, Manfrida G. Sensitivity analysis and dynamic modelling of the reinjection process in a binary cycle geothermal power plant of Larderello area. Energy. 2021;214(1):118869. DOI: 10.1016/j.energy.2020.118869

[54] Ármannsson H. CO2 emission from Geothermal Plants. In: Proceedings of International Geothermal Conference; Sept. 2003; Reykjavík, Iceland: 2003. p. 56-62

[55] Başoğul Y. Environmental assessment of a binary geothermal sourced power plant accompanied by exergy analysis. Energy Convers. Manag. 2019;195:492-501. DOI: 10.1016/j.enconman.2019.05.033

[56] Bravi M, Basosi R. Environmental impact of electricity from selected geothermal power plants in Italy. J. Clean. Prod. 2014;66(1):301-308. DOI: 10.1016/j.jclepro.2013.11.015

[57] CARBFIX project web site. [Internet]. Available: https://www.carbfix.com. [Accessed: 18-Feb-2021]

[58] Yilmaz C, Kanoglu M, Abusoglu A, Thermoeconomic cost evaluation of hydrogen production driven by binary geothermal power plant. Geothermics. 2015;57:18-25. DOI: 10.1016/j.geothermics.2015.05.005

[59] Koroneos C, Polyzakis A, Xydis G, Stylos N, Nanaki E. Exergy analysis for a proposed binary geothermal power plant in Nisyros Island, Greece. Geothermics. 2017;70:38-46. DOI: 10.1016/j.geothermics.2017.06.004

[60] Gnaifaid H, Ozcan H. Development and multiobjective optimization of an integrated flash-binary geothermal power plant with reverse osmosis desalination and absorption refrigeration for multi-generation. Geothermics. 2021;89:101949. DOI: 10.1016/j.geothermics.2020.101949

[61] Zhou C. A fundamental study on hybrid geothermal energy systems. [thesis]. University of Newcastle Australia; 2014.

Chapter 3

Quantitative Approximation of Geothermal Potential of Bakreswar Geothermal Area in Eastern India

Chiranjit Maji, Hirok Chaudhuri and Saroj Khutia

Abstract

Proper utilization of geothermal energy for power generation is still overlooked in India even after having enough potential as much as the equivalent to its other nonconventional energy resources. The source of geothermal energy is the decay of the radio-nuclei present inside the Earth's crust apart from the primordial heat source. The noble gas ^{4}He is also produced during the radioactive disintegration process. Therefore, measuring the amount of ^{4}He gas along with some other geo-chemical parameters in an Indian geothermal area, the potential of the reservoir can be evaluated. Mathematical calculations relating to the radioactive disintegration to estimate the geothermal potential of Bakreswar geothermal reservoir utilizing the concept of the ^{4}He exploration technique has been described here. The study showed that the heat (radiogenic) energy generated by the radioactive decay of ^{232}Th, ^{238}U, and ^{235}U inside the reservoir was evaluated as 38 MW. This value raises to 76 MW when primordial heat is included. The detail calculations suggest that a Kalina cycle based binary power plant using ammonia–water mixture as working fluid is supposed to be installed at the identified locations with a drilling depth of about 1,100 m and the plant would be capable of delivering the power of 9.88 MW to 40.26 MW.

Keywords: hot springs, radioactive disintegration, helium generation, geothermal power, geothermal power plant

1. Introduction

The origin of the geothermal energy is connected with the internal structure of the planet and the physiochemical processes occurring therein. According to the current knowledge, geothermal energy is unevenly distributed throughout the globe near the surface to the deep interior of the Earth [1, 2]. Depending upon the accessibility as well as the opportunities for the utilization of modern technology, many nations in the world are exploiting this natural energy resources for the commercial production of electric power [3, 4]. Geothermal energy hence, geothermal areas are generally defined through the parameter, geothermal gradient, which is the rate of the increment of the temperature profile of underneath bedrock of the Earth. The average (global) value of the geothermal gradient is typically 30 °C/km in the continental crust and 100 °C/km in the oceanic crust [1, 5]. However, in geothermal areas, its values are well above (>40 °C/km) the global average value [6]. It is so because of the magmatic intrusion. This intrusion is nothing but the molten magma, trapped within the Earth's crust at a depth of 5–10 km beneath the surface. This may still in a fluid state or the process of solidification and releasing heat constantly [2, 7, 8]. According to the origin of geothermal energy, it is categorized into two. One was from a relic of the Earth's accretion process, in which huge energy was trapped within the Earth's interior (~4.5 billion years ago) [7]. This one is named as the primordial heat source. Another one is the radiogenic heat source, which is produced by the natural decay process of long-lived radioisotopes such as ^{238}U, ^{235}U, ^{232}Th, and ^{40}K. These nuclei, of which the half-life ($T_{1/2}$) are comparable to the age of our planet, are found with significant abundance within the crust of the geothermal areas [9, 10]. A considerable amount of heat is contributed from the natural radioactive decay process. Eq. (1) to Eq. (3) represent the physicochemical processes and the heat energy released from the naturally occurring radioactive disintegration in each of the complete decay chain [11–14]. Moreover, it shows the produced crustal He (^4He) atoms and neutrinos during each decay process.

$$^{232}_{90}\mathrm{Th} \rightarrow {}^{208}_{82}\mathrm{Pb} + 6\,{}^{4}_{2}\mathrm{He} + 4\,{}^{0}_{-1}\mathrm{e} + 42.60\ \mathrm{MeV/atom} \quad (1)$$

$$^{238}_{92}\mathrm{U} \rightarrow {}^{206}_{82}\mathrm{Pb} + 8\,{}^{4}_{2}\mathrm{He} + 6\,{}^{0}_{-1}\mathrm{e} + 51.70\ \mathrm{MeV/atom} \quad (2)$$

$$^{235}_{92}\mathrm{U} \rightarrow {}^{207}_{82}\mathrm{Pb} + 7\,{}^{4}_{2}\mathrm{He} + 4\,{}^{0}_{-1}\mathrm{e} + 46.40\ \mathrm{MeV/atom} \quad (3)$$

Moreover, within the deep Earth, the production rate of He from ^{232}Th and ^{238}U [and ^{235}U] radio-nuclei are encountered to be $2.43 \times 10^{10}0\ atoms/m^3/s$ and $1.03 \times 10^8\ atoms/m^3/s$, respectively [10]. The fact of characteristics heat–helium coherence at any geothermal system (under the deep reservoir) is interpreted by such physicochemical processes [15]. This radiogenic heat, which is one of the main sources of the Earth's internal heat, powers all geodynamic processes underneath [16]. Generally, geothermal heat is transferred from the aquifer (reservoirs) to the Earth's surface by the conduction and convection process. Here, the geothermal fluid (meteoric water) acts as the carrier [1] and the radiogenic He, being highly diffusive gas, generated from the host mineral and mixes by diffusion with the fluid that circulates into the deep Earth [17]. The reservoir, which is nothing but a volume consisting of hot permeable rocks, is usually sandwiched by capping of impermeable rocks. And it is favourably connected to a recharge (surficial) area [18], from which geothermal fluids percolated to recharge the aquifer cyclically [18, 19]. The circulating fluids, to which the heat is transferred from the reservoir, escape through fracture and features from the deep reservoir and manifest through geysers, fumaroles, hot springs, etc. [8, 19]. Moreover, through diffusion and advection process, the radioactive inert gases (like ^{222}Rn & ^{220}Rn) including the stable and inert gases (such as He, Ar) are spontaneously migrating upward from the deep Earth to the superimposing atmosphere [20–22]. This process, known as 'Earth degassing', is non-uniform over space & time [23, 24]. The prominent signature of this degassing is generally noticed along active faults, fractures, oceanic ridges, geothermal fields, and even deep wells [25–27].

It is notable that geothermal energy sources are still overlooked in India for power generation even after the existence of a lot of potential resources, which are seen in twelve geothermal zones of the country [19]. However, several of them could be well utilized for the generation of power by means of developing geothermal power plants. For the sake of investigation, the hot spring site at Bakreswar in

West Bengal, India, was selected as shown in **Figure 1**. Now, knowing the amount of by-product, He gas which is ultimately reaching the surface through the fracture, fissure and hot springs vents, etc., the associated heat energy (radiogenic) produced inside the reservoir can be estimated. The energy released per unit time from underneath bedrock at the study area was calculated by means of measuring the average amount of He emanated from Agni Kunda hot spring at Bakreswar. Here mainly the decay series of ^{238}U, ^{235}U, and ^{232}Th were considered, and the amount of heat energy contributed due to each series was evaluated. Here the question may arise that each decay series [Eq. (1) to (3)] takes a long period (in geological time scale) to complete its disintegration process and release a certain amount of heat and He discretely. But, heat and He generated due to each series were utilized to calculate the amount of heat production at the said reservoir at a certain instant of time. However, He emanation at the study area shows stable activities for a long-time-interval (5 years), as established by [19]. Therefore, He generation is also stabilized for a long period, i.e., He generation due to every radioactive decay series

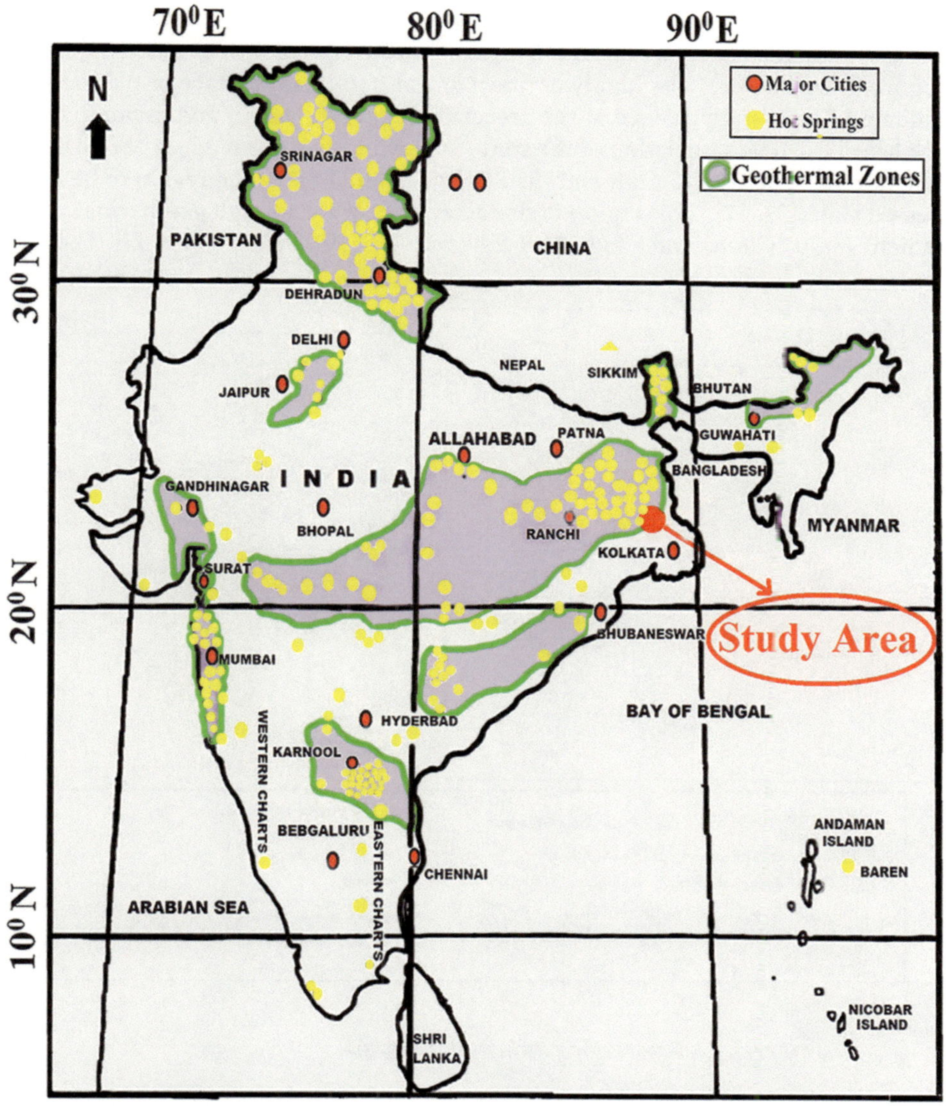

Figure 1.
Location of the study area Bakreswar in the map of India (modified after [19]).

and emanation of the said gas is in an equilibrium condition. Therefore, no He is being stored at the reservoir at the instant, and, therefore, the He emanation could be considered to be equal to the generation of the same due to the radioactive disintegration process.

2. The study area: Bakreswar geothermal province

A cluster of seven[1] hot springs is scattered over Bakreswar geothermal area within a confined zone of the surface area of about 3350 sq. m [19, 28]. The area, which is a geologically complex, heterogeneous, and extensively faulted region, is situated at the eastern end of the SONATA (Son–Narmada–Tapi) geothermal province (**Figure 2**, window a) [19, 28]. The area lies in the West Bengal Basin (WBB), the extension of the Chotanagpur Gneissic Complex [29]. Furthermore, it is linked with a 1.2-km-long shear zone, which is characteristic by 50 m wide breccia/cherty quartzite aligned through the almost north–south trend-line [26] (**Figure 2**, window b). The springs here are connected with the extinct Rajmahal volcanic activity (115 Ma), and hence, are associated with the Precambrian granitic rocks [30] (**Figure 2**, window b). The highly permeable and porous subsurface of the site is facilitated due to the presence of the brecciated, highly sheared, and mylonitized rock here [26]. The association of the study area with the eastern edges of two major fault systems (the ONGC fault and the SONATA fault) made the region to be in a stressed state [31, 32]. This region is characterized by a very high geothermal gradient (~90 °C/km) and a high heat flow rate (~ 230 mW/m^2) [19, 33]. The

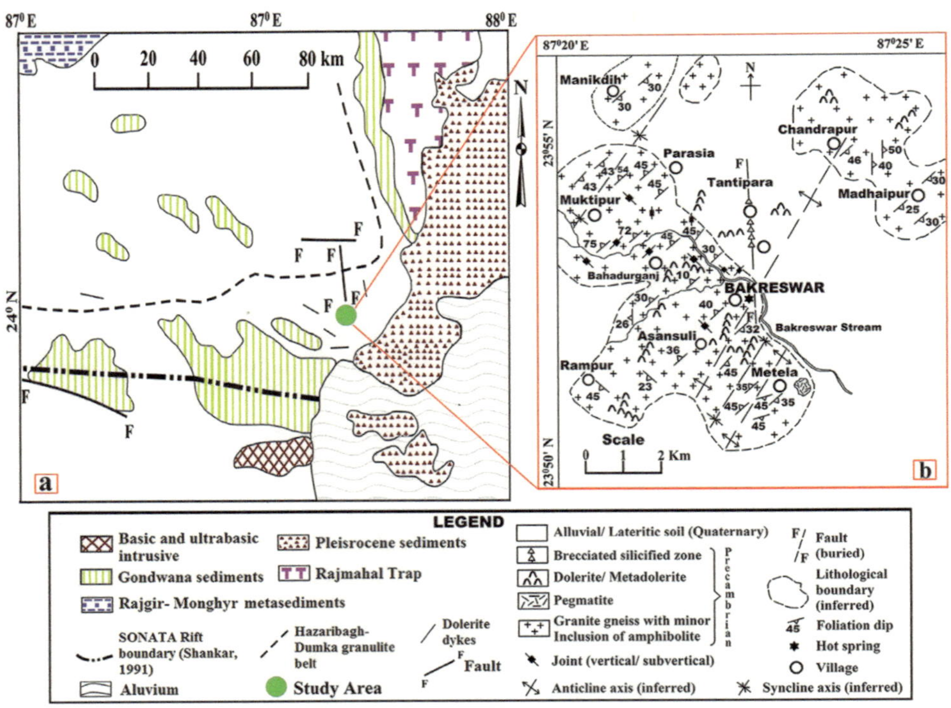

Figure 2.
(a) Regional and (b) Local geological maps of the study area (modified after [19, 26]).

[1] He emanated from other six hot springs and through the soil (soil gas) of the geothermal area are not included in the estimation due to the lack of other necessary information.

presence of a high heat-conducting zone in this area is confirmed by electrical resistivity studies. This conducting zone, which starts at a depth of around 2.8 km and goes down up to a depth of 4 km [34], is supposed to act as the heat feeder to the fault system linked with the Bakreswar hot springs. It is to be noted that the crustal thickness at the study area is only 24 km, whereas the average of the same throughout the country is 38 km [29, 35]. Besides, the average density of the crustal substance here is relatively low. Therefore, inert volatiles like He and ^{222}Rn gases can easily transmit to permeate through crustal constraints due to the presence of the thinner lithospheric overburden here. As a result, the spring and the soil gases here are dominated by the presence of high ^{222}Rn and He flux [29]. High ^{222}Rn and He gases are continuously conveyed and dispersed into the atmosphere via molecular diffusion and the formation of micro-bubbles at the hot spring vents. Here, temperature and He emanation profile of some sites of Bakreswar geothermal provinces are also tabulated in **Table 1** for a reference to attain the brief geophysical properties of the study area. Moreover, the reservoir temperature of the geothermal system underneath Bakreswar was predicted to be 100 ± 5 °C (at ~1 km depth) by [38]. The same was estimated to be in the range of 130 °C to 175 °C (by Na/K ratio) and 110 °C to 124 °C (by $TSiO_2$) by [39]. Furthermore, the range of the reservoir temperature was also evaluated as 212 °C to 124 °C, 118 °C to 120 °C, and 126 °C to 130 °C by means of silica geothermometry by [32, 38, 40] respectively. The audio-magnetotelluric (AMT) studies of the sub-surface beneath the Bakreswar geothermal area were conducted by [41] (**Figure 3**). The rapid relaxation inversion (RRI) for both transverse-electric (TE) and transverse-magnetic (TM) modes was carried out to figure out the resistivity profile of the subsurface of the site. Here, the suitable locations for drilling for the installation of a future geothermal power plant

Sl. No.	Test Site (distance from Agni Kunda)	Sample type	Temperature (°C)	He Conc. (vol %)	References
1	Bakreswar Agni Kunda (0 m)	HSG	69.0	1.72	[19]
2	Bakreswar Khar Kunda (16 m)	HSG	68.0	1.36	[36]
3	Bakreswar Bhairab Kunda (7 m)	HSG	62.0	1.12	[28]
4	Bakreswar Brahma Kunda (20 m)	HSG	46.0	1.26	[28]
5	Bakreswar Surya Kunda (18 m)	HSG	63.0	0.31	[28]
6	Bakreswar Reserve Tank (5 m)	HSG	52.0	0.91	[28]
7	PWD Bungalow at Bakreswar (987 m)	SG (1 m depth)	32.0 (Ambient)	0.35	[37]
8	PWD Bungalow at Bakreswar (988 m)	SG (3 m depth)	31.0 (Ambient)	0.02	[37]
9	PWD Bungalow at Bakreswar (990)	AA (1 m height)	33.0 (Ambient)	0.05	[29]
10	Bhabanipur (10 km)	AA (1 m height)	28.0 (Ambient)	0.07	[29]
11	Mallarpur (43 km)	BG (100 m depth)	58.0	1.20	RWA

Note: HSG = Hot spring gas; SG = Soil gas; AA = Ambient air; BG = Borehole gas; Conc. = Concentration; RWA = Recent work by the authors.

Table 1.
Temperature and He emanation profile of some sites at Bakreswar geothermal province.

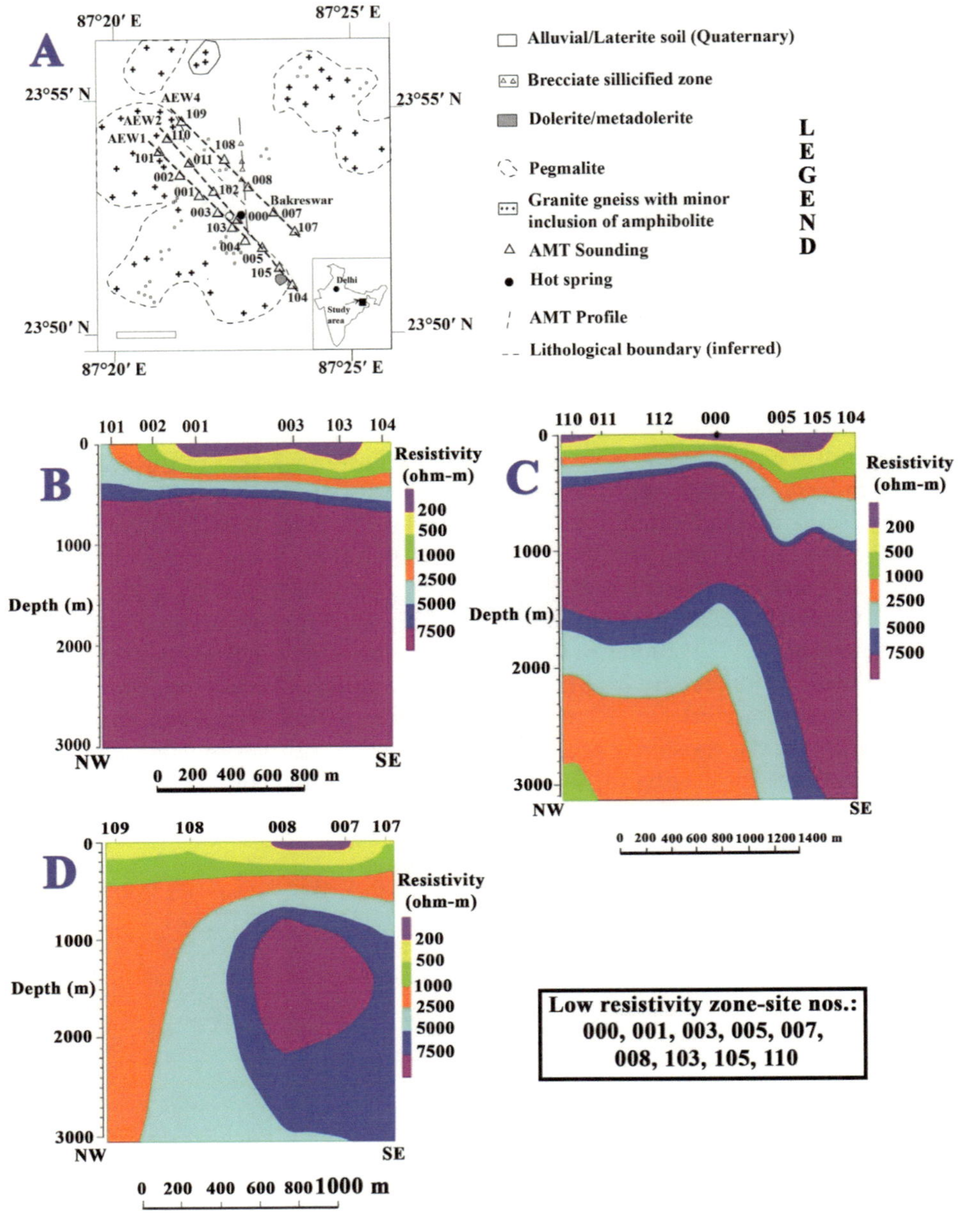

Figure 3.
(A) AEW traverses (AMT sites) on the map of the study area; (B) 2D RRI (rapid relaxation inversion) along traverse AEW1; (C) 2D RRI along traverse AEW2; (D) 2D RRI along traverse AEW4 (modified after [41] [personal communication] and [42]).

at the study area were identified by the authors using the result of that AMT survey [personal communication] and the same is discussed later.

3. Methodology

3.1 Experimental techniques

In view of continuous monitoring of gases emanated from the hot spring Agni Kunda at the spring site of Bakreswar, a field laboratory was established. In this

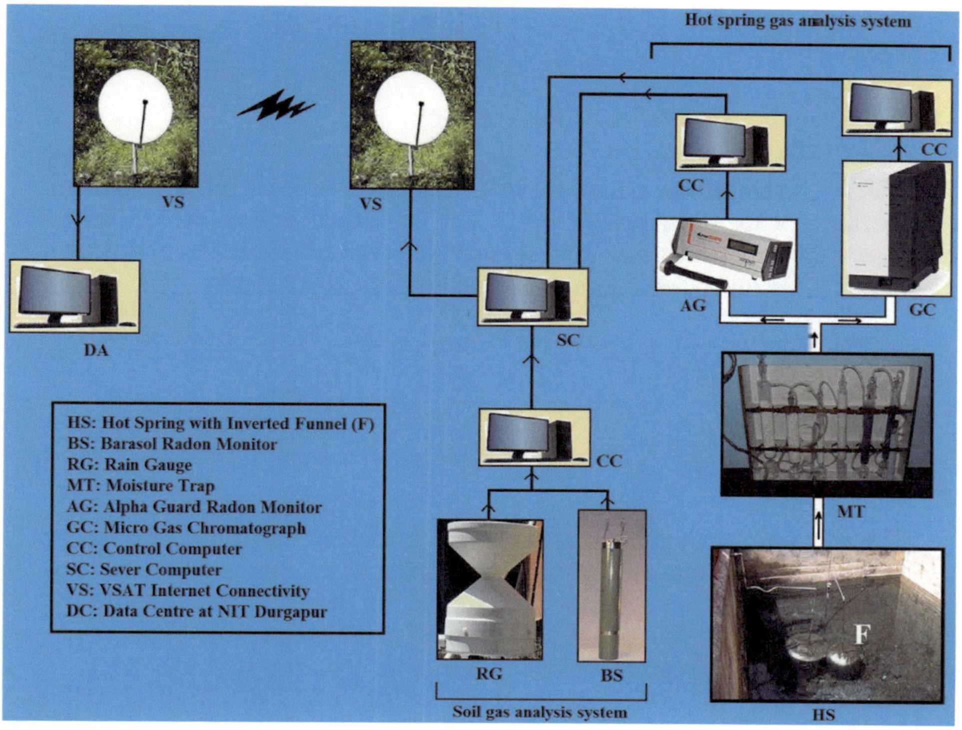

Figure 4.
Schematic diagram of the experimental set-up installed at Bakreswar (modified after [19]).

regard, a giant inverted SS funnel was placed under hot water at Agni Kunda at a position where gas out flux was significantly high, to trap hot spring gases which were comprised of He, Ar, O_2, N_2, CH_4, CO_2, ^{222}Rn, etc. A portable and programmable µ-GC (micro-gas chromatograph) CP 490 (make Agilent, Netherland) comprised of a µ-thermal conductivity detector, was utilized to detect the relative concentration of different gases present in the spring gas. Here, ultra-pure (>99.998 vol%) H_2 gas was utilized as the carrier gas for running the equipment. The entire measurement was carried out in-round-the clock (24 × 7) measurement fashion for a continuous five-year (1st August 2005 to 31st July 2010). The back-up power supply was maintained to keep a stable and continuous power supply in case of a power failure. The schematic diagram of the experimental set-up is illustrated in **Figure 4**. Further details of the above-mentioned experimental procedure are already described by [19]. Here, the average value of the He concentration (vol%) of 5 years of continuous measurement was adopted in our study. Moreover, the flow rate of the emanated gases from the spring vent was measured by means of collecting the spring gases in a gas container (5 litres) from the main channel of the incoming gas line. The gas collection procedure was kept running up to a certain time until its pressure makes equilibrium with the pressure (1.58 atm) at the spring vent underwater. This type of measurement was done once every month for a continuous five years, and the average value of those was considered as the final value of the gas flow rate under consideration. The ambient temperature of the study area was monitored for the same interval at the time of measurement of gas flow rates.

3.2 Analytical techniques

To move towards the desired direction for calculation, the following steps were adopted.

The volume of He gas (V_{He}) emanating from the spring per second was estimated as

$$V_{He} = F \times C_{He} \tag{4}$$

Where F= average flow rate of He gas emanation (recorded); C_{He}= relative concentration of He in the gas mixture, which was expelled through the spring vent (recorded). The no. of moles of He gas emanated per unit second from the hot spring was calculated using the real gas equation as stated below:

$$(V - nb)\left(P + \frac{n^2 a}{V^2}\right) = nRT \tag{5}$$

$$\text{i.e., } \frac{ab}{V^2}n^3 - \frac{a}{V}n^2 + (bP + RT)n - PV = 0 \tag{6}$$

Where V (i.e.,V_{He}) = volume of the He gas at temperature T and pressure P; R = universal gas constant = 0.0821 litre atm/ mole K; n= number of mole (to be calculated); 'a' and 'b' are real gas constants and for He, a = 0.03457 atm litre^2/mole^2; b = 0.02370 litre/mole [43]. Solving the Eq. (6) and considering the real root for 'n', the corresponding total number of He atom (N_{He}) was calculated by

$$N_{He} = n \times N_A \tag{7}$$

Here, N_A= Avogadro's number = 6.022140857 $\times$ 10^{23} [44]. The relative contribution of the individual isotope in the generation of He atoms was calculated according to their relative abundance in the natural resources because the total number of He atom is produced via the radioactive decay series of ^{238}U, ^{235}U, and ^{232}Th. Here, the same was not applicable to the 40 K series as it disintegrates through only β emission. Therefore, for production of He atoms,

$$\text{The relative contribution of Uranium } (C_U) = \frac{\text{conc. of Uranium (U)}}{\text{total conc. of Uranium (U) \& Thorium (Th)}} \tag{8}$$

$$\text{The relative contribution of Thorium } (C_{Th}) = \frac{\text{conc. of Thorium (Th)}}{\text{total conc. of Uranium (U) \& Thorium (Th)}} \tag{9}$$

The basement of the study area is predominantly composed of granite gneiss belonging to the Precambrian Chotanagpur Gneissic Complex [26, 30]. Here the relative contribution of U and Th were evaluated according to their (average) content in granite type rock material, i.e., ^{238}U [or ^{235}U] content as 4.80 ppm and ^{232}Th content as 21.50 ppm were considered [10, 45]. Moreover, natural Uranium is an admixture of ^{238}U (99.28%) and ^{235}U (0.71%) [10]. Therefore, for production of He atoms by radioactive decay,

$$\text{The relative contribution of } {}^{238}\text{U } (C_{U-238}) = \frac{99.28}{100} \times \text{C}_\text{U} \tag{10}$$

$$\text{The relative contribution of } {}^{235}\text{U } (C_{U-235}) = \frac{0.71}{100} \times \text{C}_\text{U} \tag{11}$$

The no. of the He atoms generated (in a unit second) due to the decay of radio nuclei ^{232}Th, ^{238}U and ^{235}U are respectively $C_{Th-232} \times N_{He}$, $C_{U-238} \times N_{He}$ and $C_{U-235} \times N_{He}$. According to Eq. (1) to Eq. (3), it is reflected that 6 He atoms and 42.6 MeV/atom heat energy, 8 He atoms and 51.7 MeV/atom heat energy and 7 He atoms and 46.4 MeV/atom heat energy from the decay of ^{232}Th, ^{238}U and ^{235}U are releasing respectively. Therefore, the energy release (per unit second) from ^{232}Th, ^{238}U, and ^{235}U decay can be evaluated respectively by

$$E_{Th-232} = \frac{42.6}{6} \times C_{Th-232} \times N_{He} \tag{12}$$

$$E_{U-238} = \frac{51.6}{8} \times C_{U-238} \times N_{He} \tag{13}$$

$$E_{U-235} = \frac{46.4}{7} \times C_{U-235} \times N_{He} \tag{14}$$

And the total energy generated due to the decay of all these three radioelements were

$$E_{R(Total)} = E_{Th-232} + E_{U-238} + E_{U-235} \tag{15}$$

An important issue to discuss is that the loss of generated heat energy may be considered to be negligible here as capping of the impermeable and insulating bedrock over the geothermal system prevents the heat transfer by means of conduction and convection [30, 41, 42]. Therefore, the heat energy would be stored inside the geothermal system, which may be subjected to break its dynamical stability after the accumulation of enough energy within it. However, that does not happen as excess heat is drained to the surface, along with the transfer of geothermal fluid through the spring vent [30, 42]. Moreover, here only radiogenic heat is accounted for, and the contribution of energy belonging to primordial heat sources is not included. However, [46] documented that heat from radioactive decay was contributed about half of Earth's total heat flux, and the rest was accounted for from the primordial heat source of the Earth. Considering the similar concept, we can also assume that the primordial heat source also would contribute as much as heat energy generated by radioactive decay of radio-nuclei at the reservoir of the study area.

Therefore, the heat generated by the primordial source,

$$E_{P(Total)} = E_{R(Total)} \tag{16}$$

Therefore, total energy contributed from the radiogenic and primordial source is

$$E_{Total} = E_{P(Total)} + E_{R(Total)} \tag{17}$$

Moreover, If the geothermal gradient $\left(\frac{d\theta}{dx}\right)$ is considered to be constant at least up to the depth of the reservoir (x) then the depth (x) of the reservoir could be calculated from the below stated linear relationship.

$$\theta_r = \left(\frac{d\theta}{dx}\right) \mathrm{x} + \theta_a \tag{18}$$

Where, θ_r = reservoir temperature of the geothermal system; θ_a = average ambient temperature at the study area.

4. Result and discussion

The measured parameters as well as calculated parameters such as number of He moles emanating per second (n), the total number of He atoms emanating per second (N_{He}), the contributed relative concentration of ^{232}Th (C_{Th}), ^{238}U (C_{U-238}) and ^{235}U (C_{U-235}) including the energy contributed due to decay of ^{232}Th, E_{Th-232}, ^{238}U, E_{U-238} and ^{235}U, E_{U-235} etc. are listed in **Table 2**. The Table reflects that the heat energy generated per unit second by the radioactive decay of ^{232}Th, ^{238}U and ^{235}U inside the reservoir are 31.58 MW, 6.3585 MW and 0.0467 MW respectively and together contributed as approximately 38 MW [radiogenic source, $E_{R(Total)}$]. Considering the concept of [46], as mentioned above, total heat energy [E_{Total}] related to the radiogenic source [$E_{R(Total)}$] and the primordial source [$E_{P(Total)}$] is expected to be 76 MW. Moreover, the values would be likely increased whenever the He emanations through the others hot springs (where He emanation is comparably less than that of Agni Kunda) and the vast surface area (soil gas) at Bakreswar would be included in this estimation. However, this was a little bit difficult as well as complicated due to the technical coerces and geographical constraints. It is notable that Kalina cycle based binary power plant using ammonia–water mixture as working fluid (thermal efficiency: 13–53%) [47], is already proposed to be

Sl. No.	Parameters	Parameters' value	References [Eq., if any)
1	He concentration, C_{He}	1.72 vol%	RA
2	Flow rate, F	3.5 L/min	RA
3	He emanation per minute, V_{He}	0.0602 L/min	EA [Eq. (4)]
4	Temperature inside the spring gas trapping funnel, T	342 K (69 °C)	[19]
5	Pressure inside the spring gas trapping funnel, P	1.58 atm	RA
6	Number of moles emanating per second, n	56.3877×10^{-6}	EA [Eq. (6)]
7	Total number of He atoms emanating per second, N_{He}	33.9623×10^{18}	EA [Eq. (7)]
8	The relative concentration of ^{232}Th, C_{Th}	81.7490×10^{-2}	EA [Eq. (9)]
9	The relative contribution of ^{238}U,C_{U-238}	18.1195×10^{-2}	EA [Eq. (10)]
10	The relative contribution of ^{235}U, C_{U-235}	0.1296×10^{-2}	EA [Eq. (11)]
11	Energy contributed due to decay of ^{232}Th, E_{Th-232}	31.58 MW	EA [Eq. (12)]
12	Energy contributed due to decay of ^{238}U, E_{U-238}	6.3585 MW	EA [Eq. (13)]
13	Energy contributed due to decay of ^{235}U, E_{U-235}	0.0467 MW	EA [Eq. (14)]
14	Energy accounted for radiogenic source, $E_{R(Total)}$	37.9834 MW	EA [Eq. (15)]
15	Energy accounted for primordial source, $E_{P(Total)}$	37.9834 MW	EA [Eq. (16)]
16	Total Energy accounted from radiogenic & primordial source, E_{Total}	75.9668 MW	EA [Eq. (17)]
17	Geothermal gradient, $\frac{d\theta}{dx}$	90 °C/km	[33]
18	Reservoir temperature, θ_r	130 °C	[40]
19	Average ambient temperature, θ_a	26 °C	RA
20	Depth of the geothermal reservoir, x	1155 m	EA [Eq. (18)]

Note: EA = Estimated by the authors; RA = Recorded by the authors

Table 2.
Experimental and calculated parameters.

installed at the spring site [48]. Accordingly, if such type of power plant is supposed to be installed, for say, the plant would be capable of delivering the power of 4.94 MW (minimum thermal efficiency 13%) to 20.13 MW (maximum thermal efficiency 53%) only considering the radiogenic heat source. These values are changed to 9.88 MW and 40.26 MW, respectively, when the primordial heat source is comprised of the radiogenic heat source.

Recently, the reservoir temperature [θ_r] of the Bakreswar geothermal system was estimated to be 126–130 °C by [40] by means of silica geothermometry. The average ambient temperature [θ_a] of the study area was recorded to be 26 °C. Our estimation using the mathematical relation [Eq. (18)] shows that the geothermal reservoir at the area is expected to be located at about [x=] 1,111 to 1,155 m beneath the surface. Moreover, the results of audio magnetotelluric (AMT) studies collected from [41] revealed the existence of a deep heat reservoir in the N–W part of Bakreswar. Observation sites 000, 001, 003, 005, 007, 008, 103, 105, 110, as marked in **Figure 3**, show low resistivity profile, and these are more favourable sites for deep drilling.

It is notable that no such work has been carried out to figure out the potential of Bakreswar geothermal region in terms of power harnessing capability. However, [42] estimated the geo-heat of the site to be 1158 KW-hr ($=416.88\times10^7$ Joule) by means of considering total water discharge through Agni Kunda hot spring and difference between the spring temperature and mean ambient temperature. Further, another approach may be considered here to get a comparative view in this regard. In this connection, the water discharges from the Agni Kunda and Khar Kunda hot spring were measured to be 790 L/min ($\sim$790 kg/min) and 680 L/min ($\sim$680 kg/min) respectively. The amount of the heat energy carried out (per second) by the hot water through these springs would be 10.29–10.70 MJ (equivalent to 10.29–10.70 MW in terms of power). However, the energy carried out by other springs is not included herewith due to technical difficulties. Furthermore, the energy carried out by the gaseous phase of the hot springs and the heat loss through the soil of the vast region is out of scope to be counted.

5. Conclusion

Using a simple technique by means of He exploration study at the field site, the probable energy generated inside the reservoir was estimated here. Considering the combined source of heat generation inside the reservoir system, the energy was expected to be generated from the source of power of 38 to 76 MW using the appropriate technology. The utilization of proper technology for power generation could facilitate to build a Kalina cycle based geothermal power plant (using ammonia–water mixture as working fluid) of power harnessing capability of 9.88 MW to 40.26 MW at the study area. Moreover, the values would be likely increased whenever, the He emanations through the others hot springs (where He emanation is comparably less than that of Agni kunda) and through the vast surface area (soil gas) at Bakreswar would be included in this estimation. However, this was a little bit difficult as well as complicated due to the technical coerces and geographical constraints. Furthermore, the deep drilling (production & injection well) of the proposed power plant to be rooted upto a depth of approximately 1,100 m at a location near to the hot spring area as indicated in the **Figure 3**. However, a detail geophysical survey may also be required for selecting the appropriate and exact location for drilling as well as the measurement of the horizontal (length & width-wise) and vertical (depth-wise) dimension of the geothermal reservoir at the area.

The same would be subjected to accurately calculate the possible capacity of the power plant to be installed at the site.

Acknowledgements

The authors owe a debt of gratitude to the National Institute of Technology Durgapur (NIT Durgapur) and the Ministry of Human Resource Development (Presently known as Ministry of Education), Govt. of India for providing the financial as well as Institutional support in all respect for carrying out such type of the research activities at the field site, Bakreswar.

Author details

Chiranjit Maji*, Hirok Chaudhuri and Saroj Khutia
Department of Physics, National Institute of Technology Durgapur, Durgapur, India

*Address all correspondence to: chiranjit.nitdphysics@yahoo.in; cm.15ph1102@phd.nitdgp.ac.in

References

[1] Barbier E. Geothermal energy technology and current status: an overview. Renewable and Sustainable Energy Reviews. 2002:6:3–65.

[2] Billings MP. Structural geology. New Delhi: Prentice-Hall of India Pvt. Ltd; 1990. 326 p.

[3] Breeze P. Power Generation Technologies. 3rd ed. Cambridge: Newnes; 2019.449 p. DOI:10.1016/b978-0-08-102631-1.00012-2

[4] Mburu M. Geothermal Energy Utilisation, [Internet] 2009. Available from: https://orkustofnun.is/gogn/unu-gtp-sc/UNU-GTP-SC-19-0203.pdf [Accessed: 2021-01-24]

[5] Cann JR. Metamorphism in the ocean crust. In: Talwani M, Harrison CG, Hayes D, editors. Deep drilling results in the Atlantic Ocean: ocean crust. 1st ed. Washington: American Geophysical Union; 1979. 230–238 p. DOI: 10.1029/ME002p0230

[6] Armstead HCH. Geothermal energy–its past, present and future contribution to the energy needs of man. London: E & FN Spon Press; 1983. 4 p.

[7] Csányi Ľ, Krištof V, Kušnír S, Katin M, Marci M. Geothermal Energy: Intensive Programme Renewable Energy Sources, Železná Ruda-Špičák, University of West Bohemia, Czech Republic [Internet] 2010. Available from: http://docplayer.net/26976295-Geothermal-energy-ludovit-csanyi-vladimir-kristofstanislav-kusnir-matus-katin-martin-marci.html [Accessed: 2021-01-23]

[8] Dickson MH and Fanelli M. Geothermal background. In: Dickson MH, Fanelli M, editors. Geothermal energy: utilization and technology. 2nd ed. France: UNESCO Publishing; 2003. p. 1–27.

[9] Beus AA. Geochemistry of the lithosphere. 2nd ed. Moscow: MIR Publishers; 1976. 366 p.

[10] Pereira EB. Some problems concerning the migration and distribution of He–4 and Radon–222 in the upper sediments of the crust–a theoretical model; and the development of a quadrupole ion filter for measuring He at the soil air interface [thesis]. Houston: Rice University; 1980.

[11] Cupps VR. Heavy metal clocks, U–Pb and Th–Pb dating models: radioactive dating, part 7; Acts & facts 44(5) [Internet] 2015. Available from: http://www.icr.org/article/heavy-metal-clocks-u-th-pb-dating-models/ (Accessed: 2021-01-23]

[12] Humphreys DR, Austin SA, Baumgardner JR, Snelling AA. Recently measured Helium diffusion rate for Zircon suggests inconsistency with U-Pb age for Fenton Hill Granodiorite. Transactions of the American Geophysical Union (AGU Fall Meeting Abstracts). 2003;2003:V32C-1047.

[13] Humphreys DR, Austin SA, Baumgardner JR, Snelling AA. Helium Diffusion Age of 6,000 Years Supports Accelerated Nuclear Decay Creation Research Society Quarterly 2004:41: 1–16.

[14] Faure G. Principles of Isotope Geology. 2nd ed. New York: John Wiley & Sons; 1986. 589 p.

[15] Kennedy BM, Fischer TP, Shuster DL. Heat and Helium in Geothermal Systems. In: Proceeding of Twenty–Fifth Workshop on Geothermal Reservoir Engineering (SGP–TR–165); 24–26 January 2000; California. 2000. p. 167–173

[16] Joshua EO, Alabi OO. Pattern of radiogenic heat production in rock

samples of southwestern Nigeria. Journal of Earth Sciences and Geotechnical Engineering. 2012:2(2): 25–38.

[17] Nicholson K. Geothermal Fluid. 1st ed. Berlin: Springer Verlag; 1993. 263 p. DOI: 10.1007/978-3-642-77844-5

[18] Hochstein MP. Classification and assessment of geothermal resources. In: Dickson MH, Fanelli M, editors. Small geothermal resources: a guide to development and utilization. 1st ed. New York: UNITAR/UNDP Centre on Small Energy Resources; 1990. p. 31–59.

[19] Chaudhuri H, Maji C, Seal K, Pal S, Mandal MK. Exploration of geothermal activity using time series analysis of subsurface gases data from Bakreswar hot springs area, eastern India. Arabian Journal of Geosciences. 2018:11(12):324. DOI: https://doi.org/10.1007/s12517-018-3665-5.

[20] Sathaye KJ, Smye AJ, Jordan JS, Hesse MA. Noble gases preserve history of retentive continental crust in the bravo dome natural CO_2 field, New Mexico. Earth and Planetary Science Letters. 2016:443:32–40.

[21] Etiope G, Martinelli G. Migration of carrier and trace gases in the geosphere: an overview. Physics of the Earth and Planetary Interiors. 2002:129: 185–204.

[22] Zhao X, Fritzel TLB, Quinodoz HAM, Bethke CM, Torgersen T. Controls on the distribution and isotopic composition of helium in deep ground-water flows. Geology. 1998:26(4):291–294.

[23] King CY. Gas geochemistry applied to earthquake prediction: an overview. Journal of Geophysical Research: Solid Earth.1986:91:12269–12281.

[24] Damon PE, Kulp JL. Inert gases and the evolution of the atmosphere. Geochimica et Cosmochimica Acta. 1958:13:280–292.

[25] King CY, King BS, Evans WC. Spatial radon anomalies on active faults in California. Applied Geochemistry. 1996:11:497–510. DOI: 10.1016/0883-2927(96)00003-0

[26] Nagar RK, Vishwanathan G, Sagar S, Sankaranarayanan A. Geological, geophysical and geochemical investigations in Bakreswar-Tantloi thermal field, Birbhum and Santhal Parganas Districts, West Bengal and Bihar, India. In: Pitale UL, Padhi RN, editors. Geothermal Energy in India. 1st ed. Kolkata: Geological Society of India (Special Publication 45); 1996. p. 87–98.

[27] Wakita H, Nakamura Y, Kita I, Fujii N, Notsu K. Hydrogen release: new Indicator of fault activity. Science. 1980: 210:188–190.

[28] Chaudhuri H, Sinha B, Chandrasekharam D. Helium from Geothermal Sources. In: Proceedings of World Geothermal Congress 2015; 19–25 April 2015; Melbourne. Bonn: International Geothermal Association; 2015. p. 1–14. Available from: https://www.geothermal-energy.org/pdf/IGAstandard/WGC/2015/39007.pdf [Accessed 2021-01-25]

[29] Ghose D, Chowdhury DP, Sinha B. Large-scale helium escape from earth surface around Bakreswar-Tantloi geothermal area in Birbhum District, West Bengal, and Dumka District, Jharkhand, India. Current Science. 2002: 82(8):993–996.

[30] Majumdar RK, Majumdar N, Mukherjee AL. Geoelectric investigations in Bakreswar geothermal area, West Bengal, India. Journal of Applied Geophysics. 2000:45:187–202.

[31] Majumdar N, Majumdar RK, Mukherjee AL, Bhattacharya SK and

Jani RA. Seasonal variations in the isotopes of oxygen and hydrogen in geothermal waters from Bakreswar and Tantloi, eastern India: implications for groundwater characterization. Journal of Asian Earth Sciences. 2005:25:269–278.

[32] Shanker R. Thermal and crustal structure of SONATA–azoneof mid continental rifting in Indian shield. Journal of Geological Society of India. 1991:37:211–220.

[33] Shanker R. Heat-flow of India and discussion on its geological and economic significance. Indian Minerals. 1988:42:89–110.

[34] Shalivahan S, Sinharay RK, Bhattacharya BB. Electrical conductivity structure over geothermal province of Bakreswar, Eastern India. In: Proceedings of 17th IAGA WG 1.2 Workshop on Electromagnet Induction in the Earth 2004; 18–23 October 2004; Hyderabad. Hyderabad: IAGA WG; 2005. p. 18–23.

[35] Mukhopadhyay M, Verma RK, Ashraf MH. Gravity field and structures of the Rajmahal Hills: example of the paleo-Mesozoic continental margin in eastern India. Tectonophysics. 1986:131 (3):353–367.

[36] Chaudhuri H, Ghose D, Bhandari RK, Sen P, Sinha B. The enigma of helium. Acta Geodaetica et Geophysica Hungarica. 2010:45: 452–470. DOI: 10.1556/AGeod.45. 2010.4.5

[37] Chaudhuri H, Seal K, Maji C, Pal S, Mandal MK. The unrevealed facts on helium resources of India. Arabian Journal of Geosciences. 2019:12:216. DOI: 10.1007/s12517-019-4369-1.

[38] Majumdar N, Mukherjee AL, Majumdar RK. Mixing hydrology and chemical equilibria in Bakreswar geothermal area, Eastern India. Journal of Volcanology and Geothermal Research. 2009:183:201–212.

[39] Mukhopadhyay DK, Sarolkar PB. Geochemical appraisal of Bakreshwar-tantloi hot springs, West Bengal and Jharkhand, India. In: Proceedings of 37th Workshop on Geothermal Reservoir Engineering; 30 January-01 February 2012; Stanford. Stanford: Stanford Geothermal Program; 2012. p. 1010–1014.

[40] Gupta T, Maji C, Gupta RK, Das R, Chaudhuri H. Harnessing geothermal energy at Bakreswar geothermal area. In: Chaudhury S, Hazra AK, Balachandran S, editors. Green energy and sustainable environment. 1st ed. Kolkata: Akshar Publication; 2016. p. 57–70.

[41] Sinharay RK, Srivastava S, Bhattacharya BB. Audiomagnetotelluric studies to trace the hydrological system of thermal fluid flow of Bakreswar Hot Spring, Eastern India: A case history. Geophys. 2010:75(5):B187-B195 [personal communication] DOI: https://doi.org/10.1190/1.3431532.

[42] Majumdar RK, Majumdar N, Mukherjee AL. 2010 Geological, geochemical and geoelectric studies for hydrological characterization and assessment of Bakreswar thermal springs in hard rock areas of Birbhum District, West Bengal, India. In: Proceedings of 8th Biennial International Conference and Exposition on Petroleum Geophysics; 1–3 February 2010; Hyderabad. 2010. p. 186–193.

[43] Fishbane PM, Gasiorowicz S, Thornton ST. Physics for Scientists and Engineers. 3rd ed. New York: Pearson Prentice–Hall; 2005. 240 p.

[44] Fox RF, Hill TP. An Exact Value for Avogadro's Number. American Scientist. 2017:95:104–107.

[45] Brown AA. Formation of High Helium Gases: A Guide for Explorationist. AAPG [Internet]. 2010.

Available from: http://www.searchanddiscovery.com/pdfz/documents/2010/80115brown/ndx_brown.pdf.html [Accessed 2021-01-24]

[46] Gando et al. Partial radiogenic heat model for earth revealed by Geoneutrino measurements. Nature Geoscience. 2011:4:647–651.

[47] DiPippo R. Second Law assessment of binary plants generating power from low-temperature geothermal fluids. Geothermics. 2004:33:565–586.

[48] Das R, Gupta RK, Gupta T, Maji C, Chaudhuri H. Study on Geothermal Power Generation Techniques Related to Bakreswar-Tantloi Geothermal Area. In: Proceedings of *International Conference on Renewable Energy – Extension & Outreach*; 20–21 March 2016; Santiniketan. Available from: https://www.researchgate.net/publication/319293219_STUDY_ON_GEOTHERMAL_POWER_GENERATION_TECHNIQUES_RELATED_TO_BAKRESWAR-TANTLOI_GEOTHERMAL_AREA [Accessed: 2021-01-24]

Chapter 4

Recent Progress in District Heating with Emphasis on Low-Temperature Systems

Mostafa Khosravy

Abstract

District heating plays an important role in future sustainable energy system by integrating any available heat source, including waste heat and renewable heat sources such as geothermal or solar heat. The low-temperature district heating system is the latest generation of district heating. It was introduced less than ten years ago in adaption to the need for lower heat demand of energy-efficient buildings. The low-temperature district heating system provides an infrastructure for a higher share of renewable energy sources while reduces heat loss in pipes. Several small-scale projects were commissioned since the introduction of the technology, and many existing district heating systems are in the process of adaptation. The recent progress of low-temperature district heating systems has been discussed here. First, the fundamental knowledge that is required to understand the main advantages of a low-temperature district heating system was explained briefly. Then the most recent and important projects were discussed with emphasis on solar and geothermal district heating systems. The results of case studies show that the low-temperature solution has the lowest capital costs and has a unique position to be the primary source for building heating demand.

Keywords: district heating, low-temperature, smart heat networks, microgrid, distributed generation, community energy, solar district heating, geothermal, energy storage

1. Introduction

Traditionally, the heating in residential and commercial buildings has been provided by individual systems such as furnaces and boilers. These methods were not only less efficient but also have been responsible for substantial amounts of greenhouse gases. District Heating (DH) systems are simply systems that are powered by a central heat source instead of by multiple individual heat sources for each building. By centralizing the heating in larger systems, it is possible to supply many buildings from one or more sources, such as Combined Heat and Power (CHP), Waste-to-Energy (WtE), and Renewable Energy Sources (RES). Several cities in Europe and throughout the world have begun to shift to DH systems. **Figure 1** provides the percentage of supplied energy with DH and the share of RES in the existing DH systems. In aggregated 28 European countries, there are more than 10,000 DH systems, which provide 9% of heating in the residential sector, 10% in the service sector, and

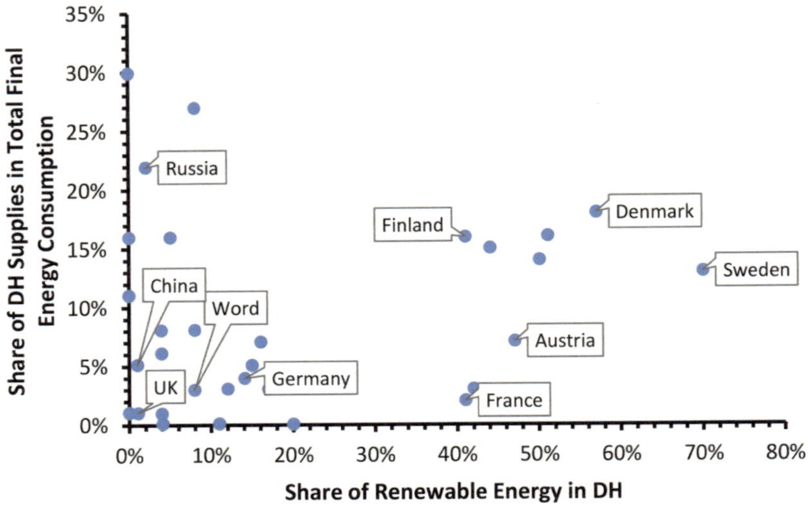

Figure 1.
Share of renewable energy in district heating networks, 2018 [1].

8% of the heating demands in the industrial sector [2]. District heating technology is less common in the United States and Canada than in Europe. According to the International District Energy Association's (IDEA) database, about 660 district energy systems are operating in the United States, and approximately 80 are working in Canada [3]. District heating is suitable for networks of all sizes, from two buildings up to a community, and even cities.

District heating systems have been evolving with a trend towards lowering supply temperatures and introducing different energy sources. Studies revealed that the 4th or 5th generation of district heating systems, along with thermal storage, is more feasible, fuel-efficient, and cheaper than individual solutions in areas with high urban density [4, 5]. The central concept of fourth-generation is a smart thermal grid. Smart thermal grids are defined as a network of pipes, connecting buildings in a neighborhood, small town, or a large metropolitan so they can be served from centralized plants or distributed heating sources, including individual contributions from the connected buildings [6]. The fifth-generation district heating has a network with temperature as close as to ambient ground temperature. In a recent review article [5], Buffa et al. studied more than forty DH systems that belong to the 5th generation. Most of these reviewed cases use shallow geothermal or groundwater as the heat source. In low-temperature networks, heat loss to the ground is eliminated, and the cost of distribution circuit is radically reduced.

If the DH supply temperature is 25°C and less, it cannot be used directly for space heating or domestic hot water (DHW). An electric heater or a booster heat pump is required to raise the temperature. A heat pump extracts heat from a low-temperature medium (e.g., DH supply) and delivers it to a medium on a higher temperature (e.g., building). In this article, a DH system with a supply temperature less than 60°C is called Low-Temperature District Heating (LTDH); thus, both 4th and 5th generations are categorized as LTDH. **Figure 2** shows the evolution of DH systems.

Several mediums can be used as heat sources of low-temperature DH. However, not all sources are universally available or have the same temperature level. Among the potential heat sources, geothermal heat was identified as the most promising source [7]. Direct use of geothermal energy in the DH system is one of the oldest and also the most common form of renewable energy. Space heating, bathing/swimming, agricultural applications, fish farming, snow melting, and industrial process are examples of direct geothermal energy utilization. Most direct uses utilize geothermal fluids in a low (30–90°C) and medium (90–150°C) temperature. The application of very low (less than 30°C) reservoir temperature has been introduced recently and initiated many types of research and case studies [8]. In a low-temperature geothermal, the thermal energy extracts from a shallow depth either by borehole heat exchangers or with the help of heat pumps. These heat pumps often are called ground-source or geothermal heat pumps (GSHP). According to WGC2020, 88 countries utilize geothermal energy for direct heat applications with significant growth in the GSHP market worldwide. About 6.46 million GSHP units have been installed in 2019, which shows a 54% increase compared to the number of installations in 2015 [9]. The trend on GSHP, as opposed to the other geothermal energy

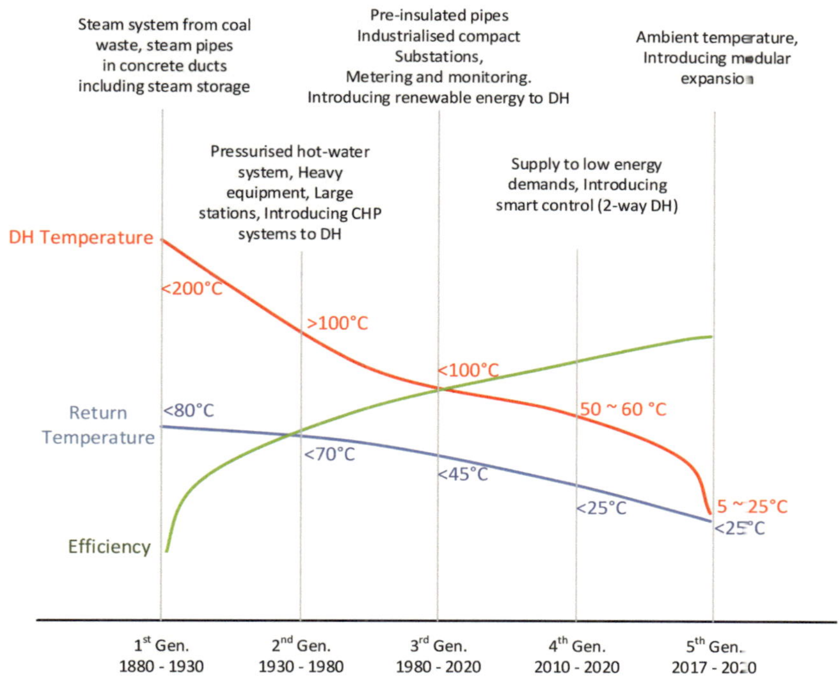

Figure 2.
Historical development of DH systems.

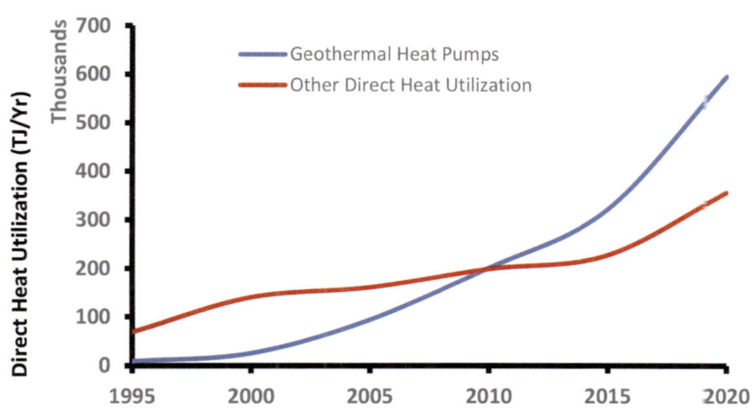

Figure 3.
Direct utilization of geothermal energy [9].

applications, has been shown in **Figure 3**. The size of installed GSHPs ranges from a couple of kilowatts for residential heating to large units over 150 kW for commercial and institutional installations. However, it is difficult to find out whether these GSHP have been installed in DH systems or not. According to the IEA, the heat supplied through DH increased by 18% in 2019 compared with 2015 [10]. Therefore, it can be concluded that most of the GSHPs were not installed in DH networks. In a comparative analysis, Lee et al. briefly discussed the advantages of DH over individual GSHP [11]. In another study, Shin et al. proposed the integration of GSHP on a shared loop to increase the system efficiency and improve the heat demand control [12].

One of the first LTDH projects with geothermal heat source is the residential area in Berlin-Zehlendorf, with 22 houses, 135 apartments, and a total of 21,000 m^2 floor space that was completed in 2016. The network temperature is approximately 10°C so that no heat is wasted, and no expensive pipe insulation is required. Decentralized heat pumps extract heat from the network and supply heat energy to the houses. Heat is provided by a CHP plant and borehole heat exchangers [13].

Another example is the city of Plymouth in the UK that will be adapted to a low-temperature DH. The supply DH temperature in the primary energy sharing network is designed between 2°C and 25°C, and the return temperature will not be higher than 25°C throughout the year. The end-users will equip with DHW booster heat pumps to increase DHW temperature to 50°C. Heat sources of the DH network will be groundwater, sea, and low-grade waste heat [14]. This particular project is one of the HeatNet pilot studies. HeatNet is an EU Interreg project to address the challenge of reducing CO2 emissions across northwest Europe by creating an integrated transnational approach to the supply of renewable and low carbon heat. The project's construction started in 2020, and the first stage of the project is planned to commission in 2021 [15].

DH systems' design requires a case-by-case approach to fully take advantage of the available local energy and identify end-users heating demand profiles. Therefore, DH systems are always site-specific and vary from one location to another, considering the size, climate, heat sources, and technologies. DH system can also be classified by size, which defines by:

- Length of the piping system (trench length)
- Number of substations
- Number of connected consumers
- Amount of investment costs
- Complexity (e.g., number of heat sources)
- Distributed heat and size of heat sources
- Spatial coverage area

Copenhagen is an excellent example of an extensive DH system. The DH system was started with one small local network in 1903, and now 98% of the city is supplied by district heating. The system serves 75 million m^2 of net floor area. The annual heat sale is 8500 GWh, and the system capacity is 10,000 GWh. The backbone of the system is a 160 km long distribution network and 3 x 24,000 m^3 heat storage tanks [16].

Small DH systems are more suitable for residential communities and small to medium size industries with excess heat. In some cases, a small DH system may connect to a large DH grid. However, the general idea is to promote individual piping networks that connect a relatively small number of consumers.

Micro heating grids are a relatively new concept characterized by advanced central control to share the resources and interact with the DH network [17]. One advantage of microgrids is that these systems could be built more straightforward and faster because of the small number of customers, without lengthy procedures.

DH systems can be categorized according to the heat production units' location into centralized and decentralized systems. Most DH systems were designed based on one of a few centralized heat generators in the past. By introducing the 4th and 5th generation, a growing number of decentralized systems use heat from various decentralized facilities. A centralized approach is best suited for upgrades or expansions of an existing district DH system. The distributed approach is recommended for a new and sparse area with relatively low load density. As such, the cost of constructing a new district energy network outweighs the other benefits of a centralized district energy system.

Nonetheless, all DH system encompasses:

- Heat sources
- Distribution network
- Consumer interconnection
- Heat storage

Despite the well-known advantages of LTDH, there are a limited number of literature reviews. The majority of the reviews only focus on a specific aspect of the district heating systems, such as modeling [18] or system flexibility [19]. In order to address the lack of a comprehensive literature review, this article provides a preliminary review of LTDH systems. The information was collected through the review of international success stories and recent academic literatures. In the following sections, the progress of low-temperature district heating systems is reviewed with respect to heat sources and distribution networks. Geothermal heat and solar radiation are the most viable types of heat source, therefore both of them are discussed in details. The cost of DH and aspects of network design are carried out by the review of typical LTDH systems.

2. Heat sources

One of the advantages of LTDH is diversified heat sources. Studies and pilot projects have shown that a DH temperature of less than 60°C significantly increases the potential to utilize waste heat of different industrial processes and cooling processes (e.g., supermarkets or data centers waste heat). Heat can be supplied by various sources such as:

- Thermal powerplants
- Waste-to-Energy (WtE) facilities (e.g., Incinerators)

- Industrial processes (i.e., Waste heat recovery system)
- Sewerage water
- Commercial buildings (e.g., Datacenter)
- Boilers (Gas fired, Electric, biomass or biogas fuelled)
- CHP plants
- Geothermal sources
- Heat pumps
- Fuel cells
- Solar thermal arrays

Heat recovery from industrial processes is not a new concept for DH. It has been applied in some countries such as Russia, Sweden, and Germany for many years in high-temperature DH networks [20]. An excellent example of the waste heat recovery is MEMPHIS's research project under IEA DHC Annex XII [21]. As part of this project, an open-source map[1] has been developed to assess waste heat potential from the industry and business sector and sewer networks. Some studies recommend adapting the industrial process heat recovery systems for LTDH [22, 23]. However, the main barrier is the economic risk associated with these heat recovery systems, if the primary industrial activities close down.

Renewable heating sources, such as solar and geothermal, are emerging in most countries. As an example, the European statistical data shows that the energy supply becomes increasingly renewable. They committed to have 100% renewable resources by 2050 [24]. A review of renewable energy sources for district heating was published recently by Olsthoorn et al. [25]. However, only the two renewable heat sources of solar and geothermal have been discussed here.

2.1 Solar district heating

Danish district heating is the most innovative district heating sector in the world. More than 1.3 million m^2 solar district heating (SDH) plants are in operation in Denmark[2]. Moreover, more than 70% of the large solar district heating plants worldwide are constructed in Denmark [26]. Since 2009, the European Union has supported three multinational SDH projects regarding solar district heating plants' market development. One of them, called "SDHp2m", addressed market uptake challenges for broader use of SDH [27]. Most of the DHp2m data are all freely available and can provide a basis for SDH feasibility evaluations.

Solar irradiance is the amount of solar radiation obtained per unit area by a given surface (W/m^2) in a location. This irradiance varies month by month, depending on the seasons. It also varies throughout the day, depending on the sun's position in the sky and the weather. The solar efficiency is the ratio between solar heat production and the total solar irradiation on the collector plane. This ratio is a performance measure on how well the system utilizes the available solar radiation. Solar efficiency mostly depends on operating conditions, such as temperature levels and solar

---

1 http://cities.ait.ac.at/uilab/udb/home/memphis/

2 http://solarheatdata.eu/

radiation intensity. Hence, low solar efficiency is not necessarily caused by a poorly working system or inefficient collectors [28]. A schematic of the SDH was shown in **Figure 4**. The monthly average solar efficiency and the total heat generated from an SDH in Vojens, Denmark were presented in **Figures 5** and **6**, respectively. This SDH commissioned in 2012 with an effective aperture area of 17,500 m^2, and a 3000 m^3 storage tank. The plant went to an expansion in 2014 and 2015, which end up with 5439 solar collectors (area of 70,000 m^2), and a thermal pit storage capacity of 200,000 m^3 for seasonal storing of excess solar heat [29].

The investment cost and the operating costs are the critical factors of the planning. The operating cost depends on the location and system components. As a rule of thumb, an annual rate of 0.54 €/MWh is considered in SDHp2m or 0.0405 €/m^2 (collector area) in the Danish Technology Data catalog[3]. It is expected that the system capital cost per MWh decreases by increasing the DH size. The capital cost is a combination of equipment costs (i.e., solar collectors, piping system, circulation pumps) and installation. **Figure 7** provides an estimation for solar collectors as per the SDHp2m study and the Danish Technology Data catalog.

Seasonal heat storage is effectively increasing solar heating in an SDH system. The ratio of heat provide by solar collectors in a typical SDH system is around 20%, if there is no seasonal heat storage [30]. The seasonal heat storage can increase the solar heating share to 30–50% [26]. Four different options of long-term or seasonal heat storage are available:

- PTES, Pit Thermal Energy Storage
- BTES, Borehole Thermal Energy Storage, ground storage with closed loops
- ATES, Aquifer Thermal Energy Storage, ground storage with open loops
- TTES, Tank Thermal Energy Storage

Pit Thermal Energy Storages (PTES) are a relatively cheap storage technology, which has been developed mostly in Denmark (e.g., Marstal 75,000 m^3, Dronninglund 60,00 m^3) in combination with solar thermal plants. The limitations and advantages of PTES briefly are shown in **Figure 8**. The physical footprint of PTES is significant; therefore, the feasibility of PTES depends on the local conditions. Borehole Thermal Energy Storage (BTES) is a relatively new technology. In a BTES, the heat directly stores underground through vertical boreholes and U-pipes. The thermal flow direction is from the center to the sides to obtain high temperatures in the center and lower at the storage boundaries during the charging period. The flow direction during the discharge is reversed. The upper surface of BTES is usually insulated to minimize the heat loss. The ground can store between 15 to 30 kWh/m^3, which is much lower than the PTES capacity of 30 to 80 kWh/m^3 [32]. The Okotoks solar district heating system that is located in Alberta, Canada, is an example of BTES. This DH system supplies more than 90% of space heating to 52 detached energy-efficient homes since 2007 [33]. An aquifer is an underground water reservoir. An Aquifer Thermal Energy Storage (ATES) utilizes a mixture of natural water and ground to store the heat. In an ATES, two wells, one warm and one cold, are drilled into the aquifer to extract and inject the groundwater. Another type of thermal storage which is very similar to PTES is called Tank Thermal Energy Storage (TTES). TTES is cylindrical steel or concrete tank placed on the ground and used daily or on a short-time storage basis. A number of guidelines and fact sheets are available through the SHC Task-45 framework. This

[3] https://ens.dk/en/our-services/projections-and-models/technology-data

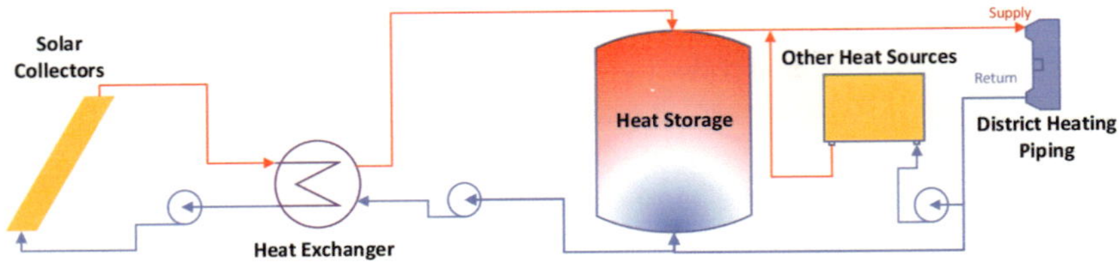

Figure 4.
Schematic of a solar DH system.

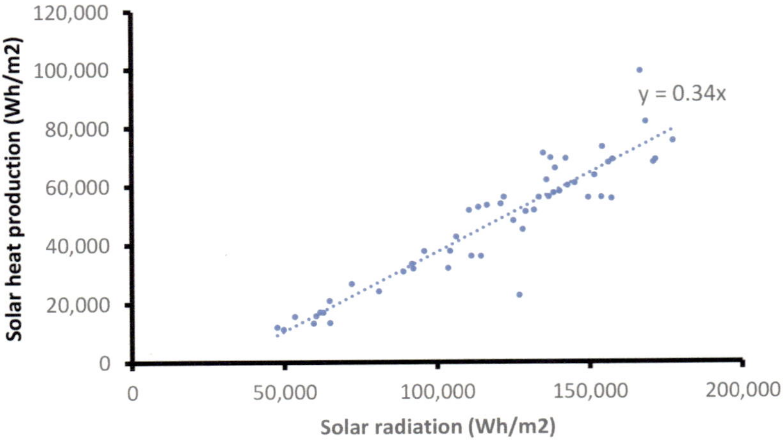

Figure 5.
Input/output plot of monthly measured values of Vojens district heating.

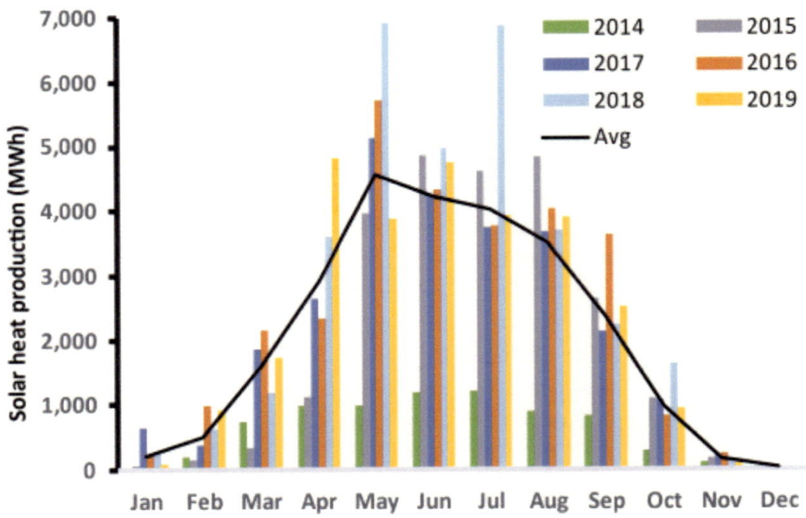

Figure 6.
Solar heat production of Vojens district heating.

framework was completed in 2014 to assist a sustainable market for large solar heating and cooling system [34].

The investment cost for design, construction, and commissioning of several European thermal storages are available (**Figure 9**). Since the design and construction of thermal energy storage systems are site-specific, **Figure 9** provides an approximate investment per storage capacity. In addition to SHC Task-45, some

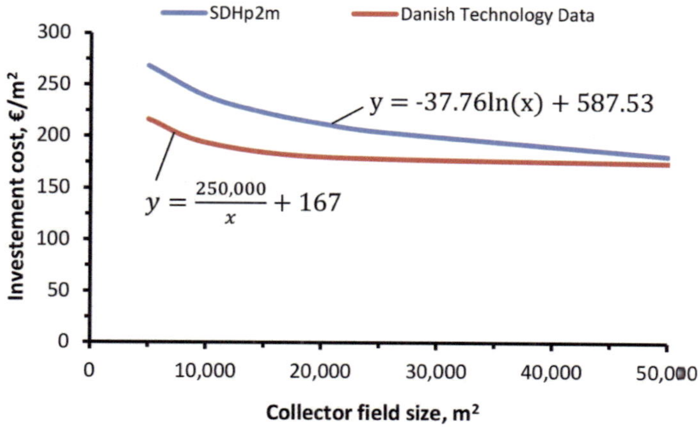

Figure 7.
Investment cost of solar collectors.

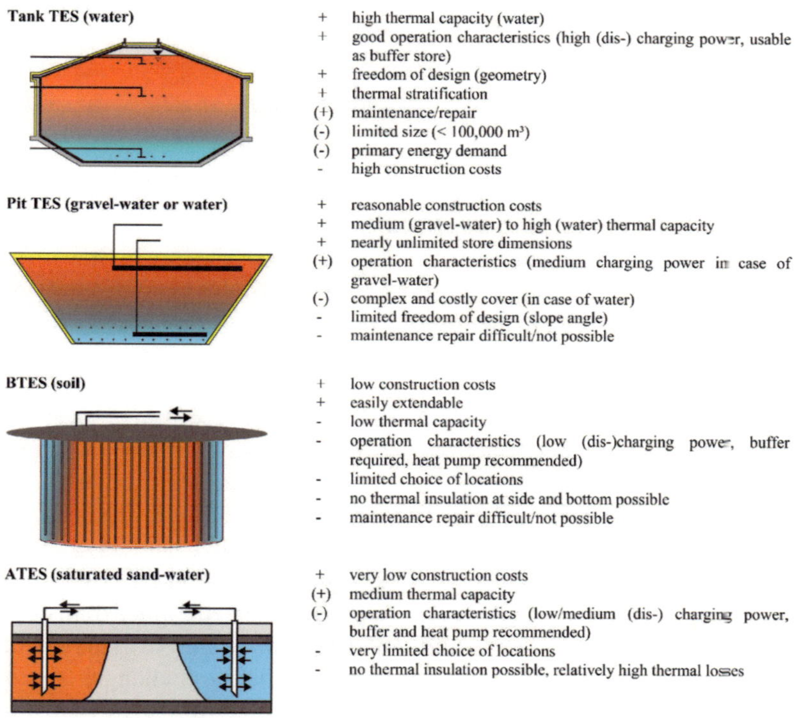

Figure 8.
Seasonal thermal energy storage concepts [31].

aspects of cost-effective largescale seasonal thermal energy storage for LTDH systems have been studied by Ochs et al. as part of the gigaTES[4] initiative [36]. However, the planning and development of seasonal thermal storage require a comprehensive study to identify the project cost.

2.2 Geothermal

Geothermal energy is a reliable and secure renewable energy source. A DH supply temperature below 60°C makes geothermal plants more advantageous to

[4] https://www.gigates.at/

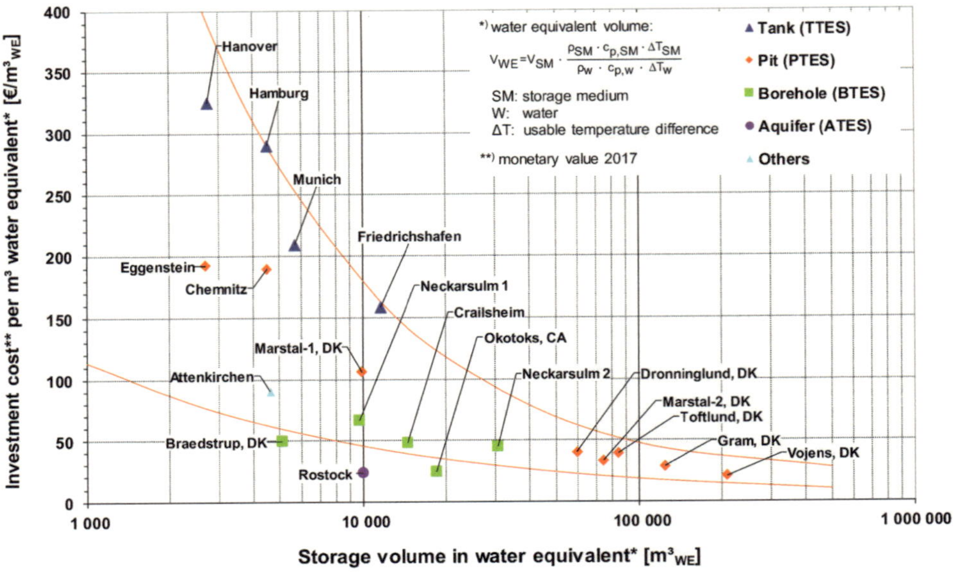

Figure 9.
*Specific investment cost of seasonal thermal energy storage [35]. ** including all necessary costs except the design and connecting pipes costs.*

satisfy the baseload in comparison with solar systems. Geothermal energy can be found independent of locations to fulfill space heating demands directly. One example of LTDH based on geothermal energy is Østre Hageby in Stavanger, Norway, where a low-temperature network with boreholes provides heating to 66 dwellings [37]. The project was completed in 2016 and reduced 61% of energy consumption for space heating and DHW (**Figure 10**).[5]

As a rule of thumb, shallow wells with temperatures between 40–150°C are suitable for hot water DH systems. In contrast, higher temperatures (deep wells) are ideal for electricity generation. The geothermal systems are capital intensive. Drilling can account for up to 50% of the total costs of a geothermal project. It has been shown that lowering the DH supply temperature reduces both the capital and operating costs of geothermal DH systems [38]. The geothermal DH system is a better option than individual geothermal heat pumps. In a comparative study in South Korea, the primary energy use of GSHP was reported higher than district heating systems [11]. Thorsteinsson and Tester have discussed the barriers of large GSHP and provided ten recommendations to overcome the challenges of geothermal district heating system development in the United States [39]. Green Energy Association has compared the DH system and individual heat pump based on the Danish data [40]. The report concludes that a new district heating system's annual operating costs are much smaller than the individual heat pump. Some case studies have been gathered in Pellegrini and Bianchini's literature review [41]. In the light of growing interest towards GSHPs, two concepts of shared GSHPs and centralized heat pumps are discussed here.

The basic principle of a GSHP is presented in **Figure 11**. Heat can be extracted from the ground at a relatively low temperature. The heated fluid is compressed to a higher pressure by a compressor. From there, a second heat exchanger or condenser transfers the heat to the home, via either warm air or circulating water.

One of the essential characteristics of GSHP is that the efficiency of the unit and the energy required to operate are directly related to the temperatures between

[5] https://www.arkitektur.no/ostre-hageby

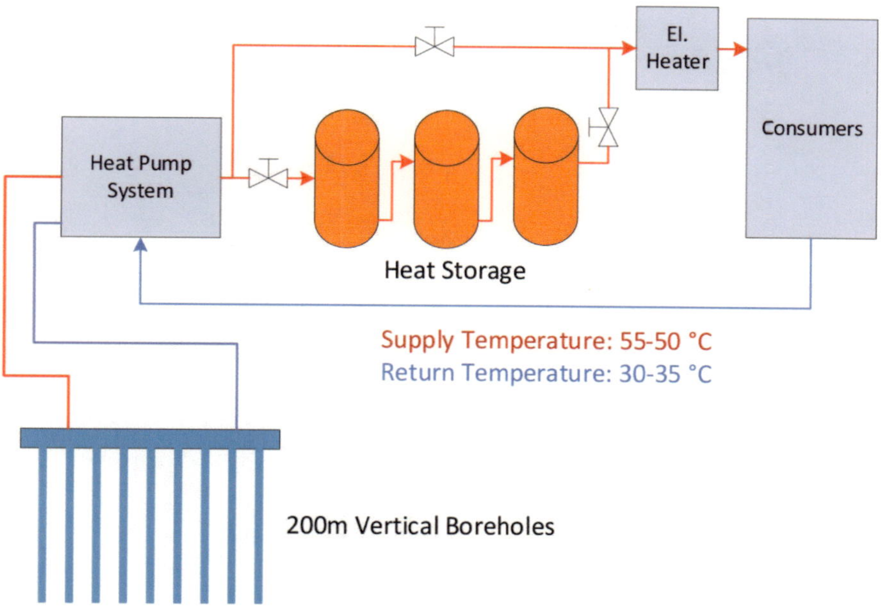

Figure 10.
Simple sketch of the Østre Hageby district heating system.

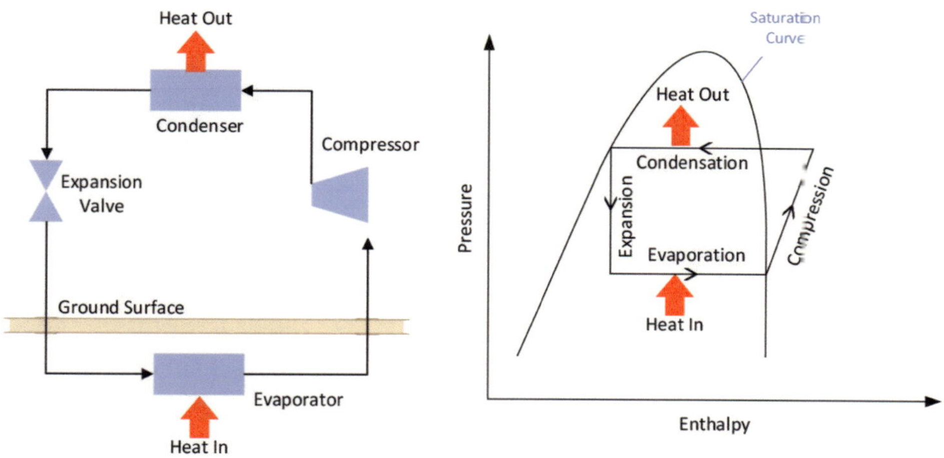

Figure 11.
Process diagram of GSHP.

which it operates. The temperature difference where the heat is absorbed (the "source") and delivered (the "sink") is called the "lift". Larger lift means greater input power to the heat pump. The heating performance of a heat pump is defined by Coefficient of Performance (COP). The COP is the heating produced divided by the energy equivalent of the electrical input resulting in a dimensionless value. The larger the COP value, the less electricity required to operate. The heat transfer between the GSHP and its surrounding soil is affected by a number of factors such as working fluid thermophysical properties and its conditions, soil thermal properties, soil moisture content, and groundwater velocity and properties.

The GSHP has excellent potential to be one of the primary energy sources in the near future. The ground energy can be tapped in a number of different ways and can be used to produce hot water as well as electricity. It has a broad spatial distribution in all countries concerning the low enthalpy resources available. Geothermal

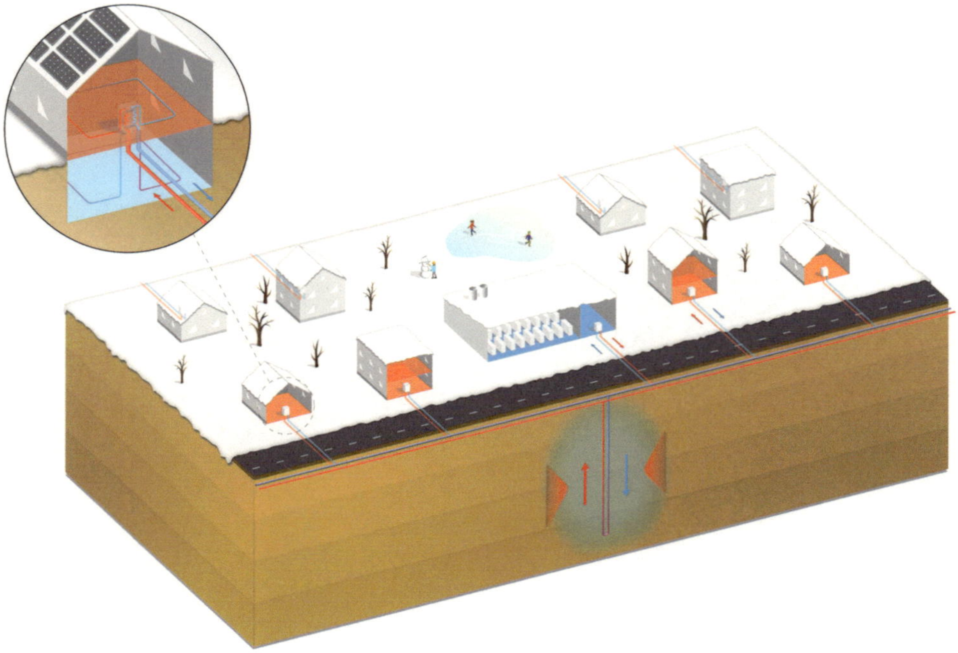

Figure 12.
Illustration of decentralized LTDH based on GSHP in a rural area of Denmark.

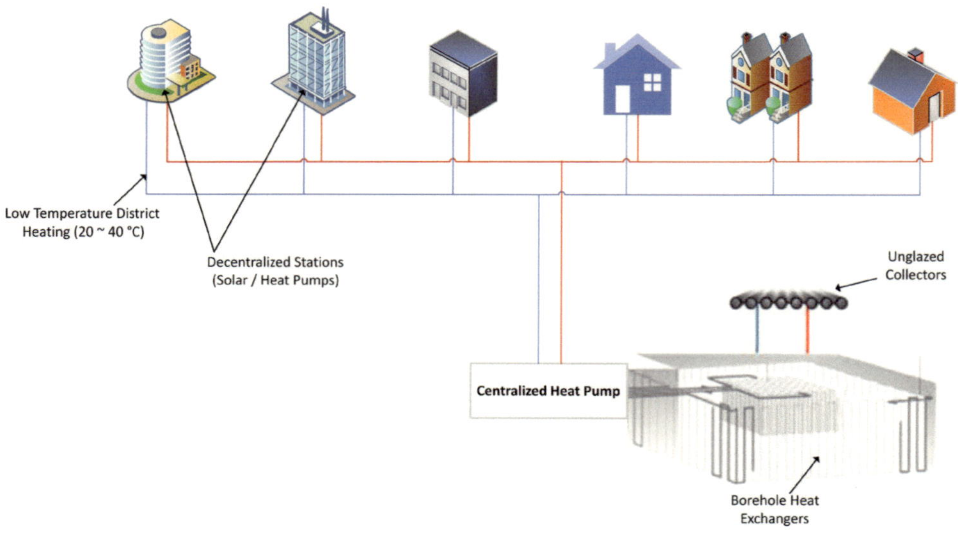

Figure 13.
District heating system of City of Kassel, Germany [46].

energy is a renewable resource that does not rely on specific factors such as the wind or the sun.

A new LTDH concept based on the use of individually adjustable and collectively managed GSHPs, connected to a low temperature non-insulated thermal distribution network has been implemented recently. This concept in the UK is called "Shared Ground Loop[6]," and the related regulations are well established [42]. In Denmark, it is called "Termonet" and is recommended mostly for rural areas. Three

[6] https://www.icax.co.uk/Shared_Ground_Loop.html

Termonet systems have been commissioned in Denmark during 2017 and 2018, with borehole heat exchangers [43] (**Figure 12**).[7]

The connected group of GSHP is based on the following ideas:

1. The aggregated heat demand can be sourced from vertical boreholes [44]

2. The investment costs reduce on the basis that a single larger ground loop will not cost as much as two separate GSHPs

3. Optimization of the heat pump (i.e., operational sequencing) will leads to more energy savings [45] and minimize the boreholes depletion.

4. Seasonal storage can be added to the system to cover both heating and cooling demands and also the distribution network can benefit from solar or other available heat sources

5. The system planning is more flexible and scalable than the centralized approach[8]

The other concept is a central shallow geothermal plant. Shallow geothermal plants have less than 400-meter deep boreholes. Since the extracted temperature can have a wide temperature range, it may require to be raised with a heat pump. **Figure 13** depicted the LTDH system of a new community with 131 low energy residential houses on a land area of 115,000 m^2, located in Kassel, Germany. The system includes a centralized GSHP with an LTDH network. Since the buildings require a higher temperature than what was provided with LTDH, especially for DHW, they were equipped with heat pumps. Different aspects of using heat pumps to balance the temperature in a district heating system have been discussed in the IEA Annex 47 project [47]. In another project (i.e., RELaTED), it has been shown that ground source heat pumps fit well with the LTDH concept [48]. However, further research is yet to come in order to fully understand all aspects of shallow geothermal energy [49].

3. Distribution network

The distribution network's role is the transmission of heat generated in centralized or distributed locations through a system of pipes for residential and commercial heating requirements. The DH heat supply must provide sufficient energy at the appropriate temperature and pressure to meet end-user demands. In LTDH planning, the system design starts with the identification of demands (**Figure 14**). The demand can be calculated on a case by case basis or estimated for a group of buildings. Several tools for heat mapping are developed in order to facilitate DH planning such as:

1. PlanHeat: http://planheat.eu/

2. THERMOS: http://www.thermos-project.eu/home

3. HotMaps: http://www.hotmaps-project.eu

[7] https://termonet.dk/

[8] https://www.kensaheatpumps.com/shoebox-ground-source-heat-pump/

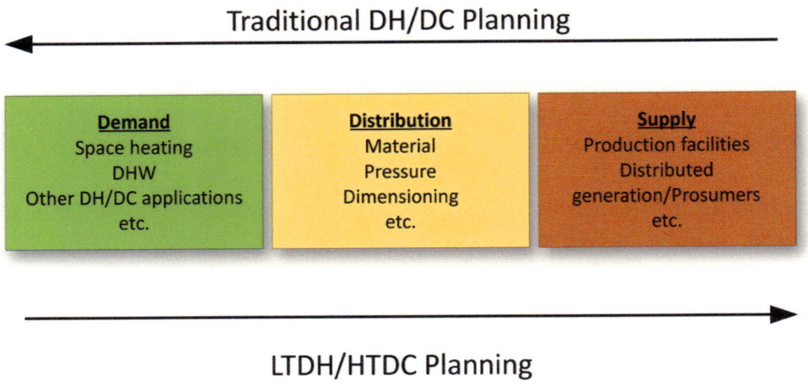

Figure 14.
Basic principles of traditional DH and LTDH planning [50].

The heating capacity of a DH network can be defined by the following three parameters, which are all related to each other.

- Spatial heat density (mean energy demand per hectare, MWh/ha),
- Specific heat density (heat demand per unit floor area in a building, kWh/m^2),
- Linear heat density (energy demand per unit length of heat network pipe, MWh/m)

After determination of heating demand, four factors influence the optimum design of an LTDH network:

1. Pipe length
2. Pipe diameter
3. Pipe insulation
4. Topography and controlling strategies of the network

Several pipe types are available, and the selection of them depends mainly on operating conditions and cost. The different kinds of pipes are ranging from rigid steel pipes to flexible plastic pipes manufacture with pre-insulated bonded. The pre-insulated flexible single or twin pipes are the standard choices for LTDH. A twin pipe integrates both the supply and return lines within one casing. Depending on the insulation thickness, both single and twin pipes are categorized in series 1, 2, or 3. The two types of single and twin pre-insulated pipes are shown in **Figure 15**.

The required pipe length is calculated by the linear length between all buildings within a hectare. Each pipe section must accommodate the peak heat loads.

The pipe diameter defines by heat density and must be carefully selected. The project capital cost and network heat loss are directly related to pipe size. In order to determine an optimum pipe diameter, different techniques have been proposed in recent publications [52–56]. Increasing the pipe size, improve the linear heat density of the DH system and, on the other hand, increases the project cost.

Linear demand density is calculated from the flow velocity and network temperature. Different guidelines recommend different thresholds for the design velocity. The impact of different design guidelines on the DH network cost has been evaluated by Best et al. [57]. They compared the guidelines of Sweden, Germany, and Austria and concluded that allowing high specific pressure drops of ≥300 Pa/m at maximum heat load leads to transportation pipe investment savings of 6–8%. An increase in the pipe diameter without additional insulation thickness increases the

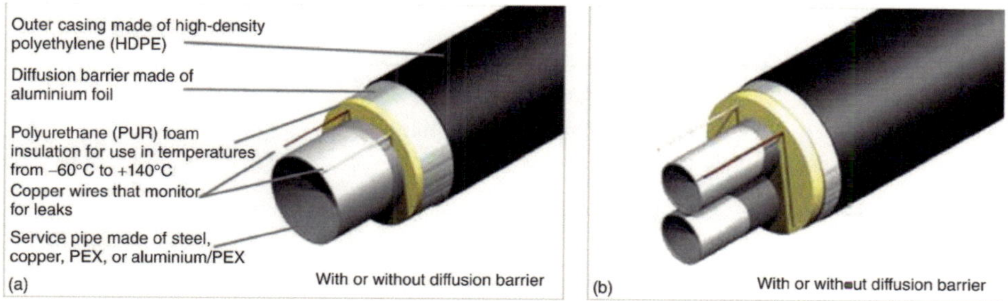

Figure 15.
Pre-insulated district heating pipe (a) single (b) twin [51].

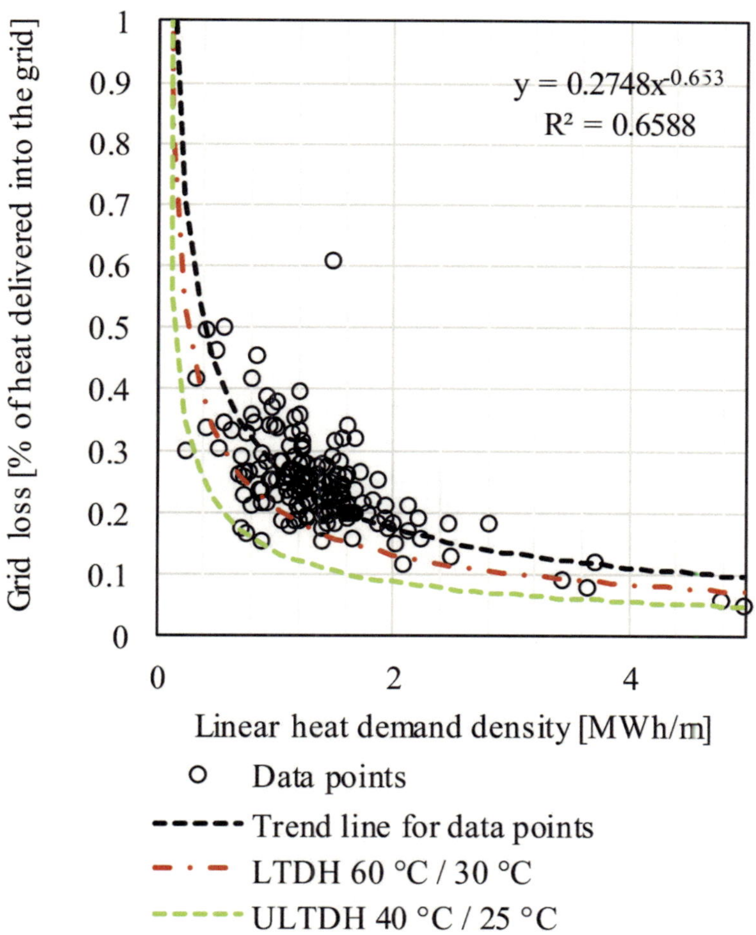

Figure 16.
Heat loss data from the existing DH networks in Denmark [58].

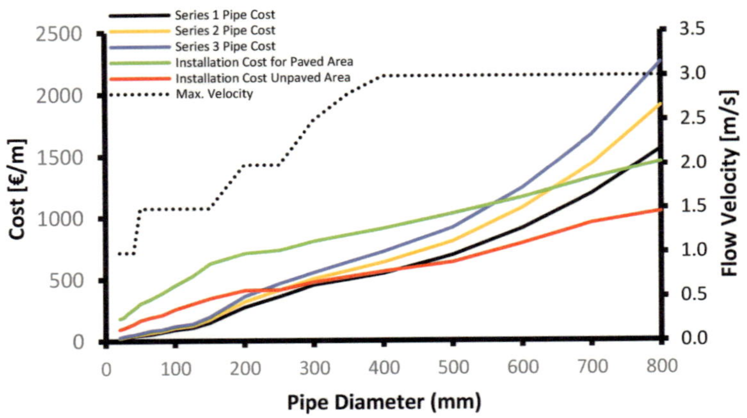

Figure 17.
Prices and maximum fluid velocity of network pipes.

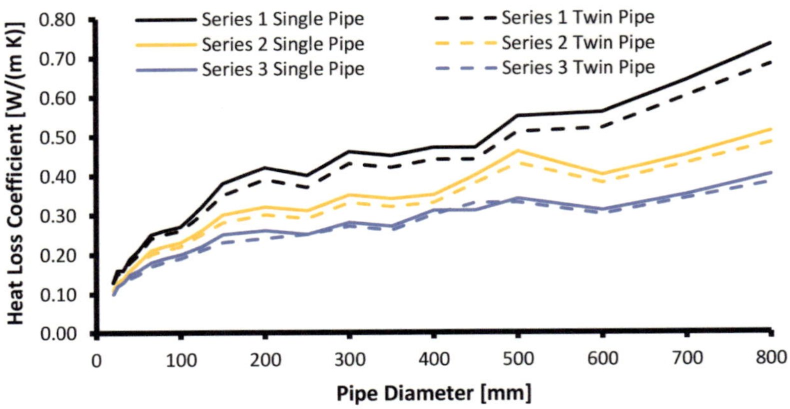

Figure 18.
Typical values of heat loss coefficients and external diameters of pre-insulated pipes.

lateral heat loss. However, this heat loss increase is minimal when compared with the heat density increase. **Figure 16** shows the magnitude of heat loss versus linear heat density.

As part of the effort to expand the LTDH networks, a research project under European Commissions, Horizon 2020, provided a useful pre-design support tool. In this project, which is called FLEXYNET[9], an Excel tool has been developed to carry out preliminary feasibility studies on the implementation of LTDH. The following cost data has been selected from this publicly available tool [59] (**Figures 17** and **18**).

The efficient operation of DH is based on complicated interactions of different heat sources with different consumers. Appropriate control of such a complex system is another challenge of LTDH systems. The control logic is a combination of head/pressure control, temperature control, and distribution optimization. Inadequate control of pressure in the DH network would lead to more water flow through the consumers close to the DH pumps and insufficient water flow through the consumers located far away. Since the supply temperature of LTDH is low, a small unpredicted variation in the demand will impact the system operation and

[9] Fifth generation, Low temperature, high EXergY district heating and cooling NETworkS: http://www.flexynets.eu/

efficiency. Some DH network design approaches are recommended to improve system flexibility. One of these solutions is the ring network [60]. Unlike the traditional designs, ring topology equalizes the pressure differences between the supply and return pipes. The rink network reduces the risk of pressure spike in case of malfunctioning of any control valves.

Further review of different controlling strategies has been provided by Vandermeulen et al. [61]. One of the recent efforts to improve the DH controls is the STORM initiative. In this project, a new control algorithm has been developed and successfully applied in two demo sites [62, 63].

4. Summary

This article described the state of the art of several existing LTDH systems. The advantages of LTDH networks over the traditional district heating networks have been discussed. Reviewed cases and studies intensified the energy efficiency potential of LTDH. This system provides a unique opportunity to integrate renewable heat sources such as geothermal and solar as much as possible. The capital costs of LTDH is generally less than the high-temperature DH. In a very low-temperature district heating concept, heat pumps are required mainly for domestic hot water at the end-users. Since the LTDH technology is relatively new, the regulations, including policies and incentives and design standards, are not well established. Further investigations are needed in order to identify the design criteria and develop regulations, including the transition from the exiting high-temperature DH to LTDH. The behavior of the LTDH under different operating conditions and various design configurations should define by holistic models that are available yet. The LTDH system has the potential to receive full market attention as the technology evolves.

Author details

Mostafa Khosravy
CLEAResult, Toronto, Canada

References

[1] How can district heating help decarbonise the heat sector by 2024? : IEA; 2019 [Available from: https://www.iea.org/data-and-statistics/charts/share-of-renewable-energy-in-district-heating-networks-2018].

[2] "Heating and cooling facts and figures," in "Heat Roadmap Europe," Heat Roadmap Europe 4 (HRE4), 2017. [Online]. Available: https://heatroadmap.eu/heating-and-cooling-energy-demand-profiles/

[3] District Energy in North America - International District Energy Association [Available from: https://www.districtenergy.org/resources/resources/system-maps].

[4] H. Lund *et al.*, "The status of 4th generation district heating: Research and results," *Energy,* vol. 164, pp. 147-159, 2018.

[5] S. Buffa, M. Cozzini, M. D'Antoni, M. Baratieri, and R. Fedrizzi, "5th generation district heating and cooling systems: A review of existing cases in Europe," *Renewable and Sustainable Energy Reviews,* vol. 104, pp. 504-522, 2019.

[6] H. Lund *et al.*, "4th Generation District Heating (4GDH). Integrating smart thermal grids into future sustainable energy systems," *Energy,* vol. 68, pp. 1-11, 2014, doi: 10.1016/j.energy.2014.02.089.

[7] "Heat Pump Driven District Heating Systems," in "District Heating Systems Solution Booklet," Smart Cities Information System (SCIS), October 2019. [Online]. Available: https://smartcities-infosystem.eu/content/heat-pump-driven-district-heating-systems

[8] D. Schmidt *et al.*, "Low temperature district heating for future energy systems," *Energy Procedia,* vol. 116, pp. 26-38, 2017.

[9] J. W. Lund and A. N. Toth, "Direct Utilization of Geothermal Energy 2020 Worldwide Review," in *World Geothermal Congress*, Reykjavik, Iceland, 2020. [Online]. Available: https://www.geothermal-energy.org/pdf/IGAstandard/WGC/2020/01018.pdf

[10] Heat supplied through DHD and % of renewables, 2007-2024: IEA, Paris; [Available from: https://www.iea.org/data-and-statistics/charts/heat-supplied-through-dhd-and-of-renewables-2007-2024].

[11] J. S. Lee, H. C. Kim, and S. Y. Im, "Comparative Analysis between District Heating and Geothermal Heat Pump System," 2017, vol. 116: Elsevier Ltd, pp. 403-406, doi: 10.1016/j.egypro.2017.05.087

[12] J. S. Shin, J. W. Park, and S. H. Kim, "Measurement and Verification of Integrated Ground Source Heat Pumps on a Shared Ground Loop," *Energies,* vol. 13, no. 7, pp. 1752-1752, 2020, doi: 10.3390/en13071752.

[13] Berlin-Zehlendorf 900 kW [Available from: https://www.geo-en.de/images/pdf/pdf/en/referenzen_en.pdf].

[14] "District Energy Plymouth- Part 2: a technical guide for designers of building services," in "Connecting to the Plymouth Heat Network," Plymouth City Council, 2019. [Online]. Available: https://www.plymouth.gov.uk/districtenergy

[15] "HEAT NETWORKS: 2019 Q2 PIPELINE," 2019. [Online]. Available: https://assets.publishing.service.gov.uk/government/uploads/system/uploads/attachment_data/

file/823506/2019_Q2_Heat_Networks_Project_Pipeline.pdf

[16] State of Green, "DISTRICT ENERGY - Green heating and cooling for urban areas," 2020, issue May 2020. [Online]. Available: https://stateofgreen.com/en/publications/district-energy/

[17] V. Olgyay, S. Coan, B. Webster, and W. Livingood, "Connected Communities: A Multi-Building Energy Management Approach," 2020. [Online]. Available: www.nrel.gov/publications.

[18] B. Talebi, P. A. Mirzaei, A. Bastani, and F. Haghighat, "A Review of District Heating Systems: Modeling and Optimization," (in English), *Frontiers in Built Environment,* Review vol. 2, no. 22, 2016-October-04 2016, doi: 10.3389/fbuil.2016.00022.

[19] Z. Ma, A. Knotzer, J. D. Billanes, and B. N. Jørgensen, "A literature review of energy flexibility in district heating with a survey of the 'stakeholders' participation," *Renewable and Sustainable Energy Reviews,* vol. 123, p. 109750.

[20] S. Werner, "International review of district heating and cooling," *Energy,* vol. 137, pp. 617-631, 2017.

[21] J. Pelda, S. Holler -Hawk, R. Geyer, R. Stollnberger, E. Gebetsroither-Geringer -Ait, and C. Sinclair -Bre, "Methodology to Evaluate and Map the Potential of Waste Heat from Industry, Service Sector and Sewage Water by Using Internationally Available Open Data," 2019. [Online]. Available: https://www.iea-dhc.org/fileadmin/documents/Annex_XII/IEA_DHC_MEMPHIS_final_Report_2019.pdf

[22] E. Wheatcroft, H. Wynn, K. Lygnerud, G. Bonvicini, and D. Leonte, "The Role of Low Temperature Waste Heat Recovery in Achieving 2050 Goals: A Policy Positioning Paper,' *Energies,* vol. 13, no. 8, pp. 2107-2107, 2020, doi: 10.3390/en13082107.

[23] H. Fang, J. Xia, K. Zhu, Y. Su, and Y. Jiang, "Industrial waste heat utilization for low temperature district heating," *Energy Policy,* vol. 62, pp. 236-246, 2013, doi: 10.1016/j.enpol.2013.06.104.

[24] T. European, H. Innovation Platform on Renewable, and Cooling, "2050 vision for 100% renewable heating and cooling in Europe," 2019. [Online]. Available: www.rhc-platform.org

[25] D. Olsthoorn, F. Haghighat, and P. A. Mirzaei, "Integration of storage and renewable energy into district heating systems: A review of modelling and optimization," *Solar Energy,* vol. 136, pp. 49-64, 2016.

[26] Z. Tian *et al.*, "Large-scale solar district heating plants in Danish smart thermal grid: Developments and recent trends," *Energy conversion and management,* vol. 189, pp. 67-80, 2019.

[27] Knowledge database – Solar District Heating [Available from: https://www.solar-district-heating.eu/en/knowledge-database/].

[28] C. Rockenbaugh *et al.*, ' High Performance Flat Plate Solar Thermal Collector Evaluation," National Renewable Energy Lab.(NREL), Golden, CO (United States), 2016.

[29] D. Fjernvarme, "Solar District Heating Inspiration and Experiences from Denmark," IEA SHC Task 55.

[30] K. S. C. S. S. Trier Daniel, "Potentials for Ground-Mounted SDH in Europe," in *5th International Solar District Heating Conference*, Graz, Austria, 2018, pp. 230-234. [Online]. Available: https://www.solar-district-heating.eu/wp-content/

uploads/2018/06/Proceedings-SDH-2018.pdf

[31] A. Dahash, F. Ochs, M. B. Janetti, and W. Streicher, "Advances in seasonal thermal energy storage for solar district heating applications: A critical review on large-scale hot-water tank and pit thermal energy storage systems," *Applied Energy,* vol. 239, pp. 296-315, 2019.

[32] D. Mangold and L. Deschaintre, "Task 45 Large Systems Seasonal thermal energy storage," 2015. [Online]. Available: https://task45.iea-shc.org/fact-sheets

[33] B. Sibbitt *et al.*, "The performance of a high solar fraction seasonal storage district heating system - Five years of operation," 2012/01//, vol. 30: Elsevier Ltd, pp. 856-865, doi: 10.1016/j.egypro.2012.11.097

[34] IEA SHC || Task 45 [Available from: https://task45.iea-shc.org/].

[35] T. Pauschinger *et al.*, "Integrated Cost-effective Large-scale Thermal Energy Storage for Smart District Heating and Cooling." [Online]. Available: www.iea-dhc.org

[36] F. Ochs, A. Dahash, A. Tosatto, and M. Bianchi Janetti, "Techno-economic planning and construction of cost-effective large-scale hot water thermal energy storage for Renewable District heating systems," *Renewable Energy,* vol. 150, pp. 1165-1177, 2020, doi: 10.1016/j.renene.2019.11.017.

[37] P. Breuhaus, "4th generation heating system using geothermal energy as the main source," Copenhagen, Denmark, 2017: 4DH Research Centre.

[38] H. Averfalk and S. Werner, "Economic benefits of fourth generation district heating," *Energy,* vol. 193, 2020, doi: 10.1016/j.energy.2019.116727.

[39] H. H. Thorsteinsson and J. W. Tester, "Barriers and enablers to geothermal district heating system development in the United States," *Energy Policy,* vol. 38, no. 2, pp. 803-813, 2010, doi: 10.1016/j.enpol.2009.10.025.

[40] C. H. Hansen and O. Gudmundsson, "The competitiveness of district heating compared to individual heating," 2018. [Available from: https://www.euroheat.org/publications/competitiveness-district-heating-compared-individual-heating]

[41] *M. pellegrini* and A. Bianchini, "The innovative concept of cold district heating networks: a literature review," *Energies,* vol. 11, no. 1, p. 236, 2018.

[42] The Renewable Heat Incentive Scheme Regulations 2018 [Available from: https://www.legislation.gov.uk/ukdsi/2018/9780111166734/regulation/11?view=plain].

[43] S. E. Poulsen *et al.*, "Geothermal Energy Use, Country Update for Denmark," Den Haag, Netherlands, 2019.

[44] K. Filonenko, K. Arendt, M. Jradi, S. Andersen, and C. Veje, "Modeling and Simulation of a Heating Mini-Grid for a Block of Buildings," 2020/03//, vol. 16: IBPSA, pp. 1971-1978, doi: 10.26868/25222708.2019.210870

[45] J. S. Shin, S. H. Kim, and J. W. Park, "Economic analysis of integrated ground source heat pumps on a shared ground loop," *Energies,* vol. 13, no. 11, pp. 2928-2928, 2020, doi: 10.3390/en13112928.

[46] D. Schmidt *et al.*, "Development of an Innovative Low Temperature Heat Supply Concept for a New Housing Area," *Energy Procedia 2017*, vol. 116: Elsevier Ltd, pp. 39-47, doi: 10.1016/j.egypro.2017.05.053

[47] "Heat Pumps in District Heating and Cooling systems - Annex 47," Sweden, 2019. [Online]. Available: https://heatpumpingtechnologies.org/annex47/final-report-for-hpt-tcp-annex-47-heat-pumps-in-district-heating-and-cooling-systems/

[48] R. Garay, "RELaTED D2.1 - Low Temperature Concepts," 2018. [Online]. Available: http://www.relatedproject.eu/related-project-low-temperature-concepts/

[49] K. P. Tsagarakis, "Shallow geothermal energy under the microscope: Social, economic, and institutional aspects," *Renewable Energy*, vol. 147, pp. 2801-2808, 2020, doi: 10.1016/j.renene.2019.01.004.

[50] "Strategies and concepts for LTDH and HTDC," CELSIUS Project, 2020. [Online]. Available: https://celsiuscity.eu/wp-content/uploads/2020/02/66X-Strategies-and-concepts-for-LTDH-and-HTDC.pdf

[51] H. Li, "Energy-Efficient District Heating." Chichester, UK: John Wiley & Sons, Ltd, 2015, pp. 1-24.

[52] J. Maria Jebamalai, K. Marlein, J. Laverge, L. Vandevelde, and M. van den Broek, "An automated GIS-based planning and design tool for district heating: Scenarios for a Dutch city," *Energy*, vol. 183, pp. 487-496, 2019, doi: 10.1016/j.energy.2019.06.111.

[53] H. I. Tol and S. Svendsen, "Improving the dimensioning of piping networks and network layouts in low-energy district heating systems connected to low-energy buildings: A case study in Roskilde, Denmark," *Energy*, vol. 38, no. 1, pp. 276-290, 2012, doi: 10.1016/j.energy.2011.12.002.

[54] P. Harney, D. Gartland, and F. Murphy, "Determining the optimum low-temperature district heating network design for a secondary network supplying a low-energy-use apartment block in Ireland," *Energy*, vol. 192, pp. 116595-116595, 2020, doi: 10.1016/j.energy.2019.116595.

[55] P. K. Olsen, C. H. Christiansen, M. Hofmeister, S. Svendsen, J.-E. Thorsen, and O. Gudmundsson, "Guidelines for low-temperature district heating," *EUDP–DEA. Denmark,* 2014.

[56] J. Chambers, K. Narula, M. Sulzer, and M. K. Patel, "Mapping district heating potential under evolving thermal demand scenarios and technologies: A case study for Switzerland," *Energy*, vol. 176, pp. 682-692, 2019, doi: 10.1016/j.energy.2019.04.044.

[57] I. Best, J. Orozaliev, and K. Vajen, "Impact of different design guidelines on the total distribution costs of 4th generation district heating networks," *Energy Procedia,* vol. 149, pp. 151-160, 2018.

[58] W. Meesenburg, T. Ommen, J. E. Thorsen, and B. Elmegaard, "Economic feasibility of ultra-low temperature district heating systems in newly built areas supplied by renewable energy," *Energy*, vol. 191, pp. 116496-116496, 2020, doi: 10.1016/j.energy.2019.116496.

[59] M. Cozzini *et al.*, "FLEXYNETS Guide Book on Fifth generation, low temperature, high exergy district heating and cooling networks," 2018.

[60] T. Laajalehto, M. Kuosa, T. Mäkilä, M. Lampinen, and R. Lahdelma, "Energy efficiency improvements utilising mass flow control and a ring topology in a district heating network," *Applied Thermal Engineering*, vol. 69, no. 1-2, pp. 86-95, 2014, doi: 10.1016/j.applthermaleng.2014.04.041.

[61] A. Vandermeulen, B. van der Heijde, and L. Helsen, "Controlling district

heating and cooling networks to unlock flexibility: A review," *Energy,* vol. 151, pp. 103-115, 2018.

[62] D. Vanhoudt, B. Claessens, J. Desmedt, and C. Johansson, "Status of the horizon 2020 storm project," *Energy Procedia,* vol. 116, pp. 170-179, 2017.

[63] T. Van Oevelen, D. Vanhoudt, C. Johansson, and E. Smulders, "Testing and performance evaluation of the STORM controller in two demonstration sites," *Energy,* vol. 197, pp. 117177-117177, 2020, doi: 10.1016/j.energy.2020.117177.

Chapter 5

An Approach for Estimating Geothermal Reservoir Productivity under Access Limitations Associated with Snowy and Mountainous Prospects

Mitsuo Matsumoto

Abstract

This chapter describes an approach to estimate reservoir productivity during the active exploration and development of a geothermal prospect. This approach allows a reservoir model to be updated by overcoming the severe time limitations associated with accessing sites for drilling and well testing under snowy and mountainous conditions. Performed in parallel with the conventional standard approach, the new approach enables us to obtain a first estimate of the reservoir productivity at an early time and to make successful project management decisions. Assuming a practical geothermal field, the procedures of the new approach are demonstrated here in detail. Finally, frequency distributions for the expected production rates and changes in the reservoir pressure at an arbitrary time are obtained during an assumed operational period.

Keywords: reservoir modeling, wellbore flow modeling, well testing, reservoir engineering, production engineering, project management

1. Introduction

The exploration and development lead time of a geothermal prospect directly affects its profitability because of the yearly interest factored into the cost. Sufficient profits purely produced by geothermal resources without financial support from other budgets are essential to enhance the development of environmentally friendly geothermal resources. This is why we cannot avoid trade-offs between exhaustively studying a geothermal reservoir and rapidly advancing a geothermal project.

Figure 1 shows an example of a fiscal annual schedule during an explorational and developmental project in which the author was involved as a reservoir engineer. The project was conducted in a snowy and mountainous area in northeastern Japan, where several geothermal projects have been conducted over the last decade [1]. In such areas, the schedules of drilling, well testing, and any other work at a site are

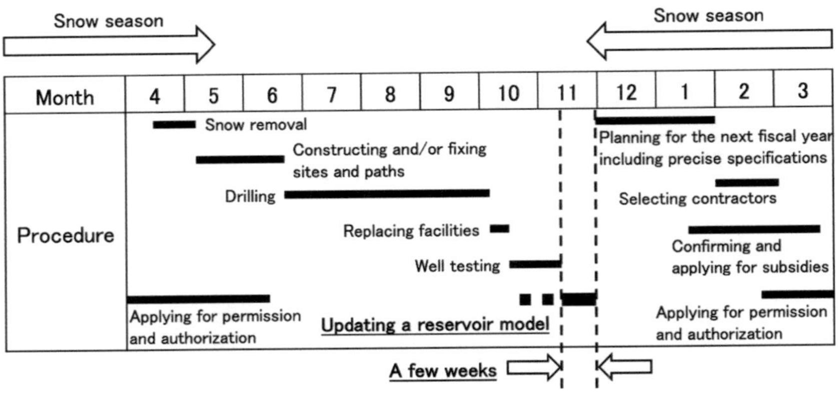

Figure 1.
Example of an annual schedule during an explorational and developmental project.

strictly limited by the snowy season. In addition, mountainous conditions limit the number of site locations that can satisfy the following conditions:

1. Accessibility to targets using directional or vertical drilling;

2. Sufficient space to install facilities for drilling and well testing;

3. Sufficient water supplies from nearby streams;

4. Accessibility to the site via paths constructed within realistic time and cost constraints including snow removal; and

5. The possibility of receiving permission and authorization while obeying numerous national regulations.

Under these severe limitations, project personnel usually identify a small number of possible locations following a large amount of effort, rather than easily selecting a location from a large number of options.

Geoscientists and engineers have only a few weeks to update a reservoir model by analyzing, considering, discussing, and updating their understanding of a geothermal system after collecting all the new data from a given year (**Figure 1**). Under such severe conditions, it is essential to account for the updated reservoir model when planning for the subsequent fiscal year. This chapter describes a concept and techniques to construct and update a reservoir model, as well as to estimate the reservoir productivity, by making the most of the highly limited time available during an active explorational and developmental project. The concept and techniques are based on the author's experience as a reservoir engineer in a real project even though specific information regarding the project, including the observational data, cannot be shown because of confidentiality reasons. The techniques described in this chapter have been partially reviewed and published in several articles and proceedings. This chapter focuses on practical procedures to construct a reservoir model by assembling these techniques.

2. Approaches to reservoir modeling

2.1 Standard approach

Let us begin by discussing the fundamental concept involved in conventional reservoir modeling approaches. As widely accepted and detailed in textbooks [2, 3], the standard legitimate approach first involves the development of a conceptual model driven by several geological, hydrological, geophysical, and geochemical observations. This first step of the standard approach establishes the basis of the understanding of a geothermal system and requires exhaustive discussions that can comprehensively and consistently explain all the geoscientific observations. A good conceptual model plays a key role in successful reservoir modeling.

After the exhaustive study needed to construct a conceptual model, a natural-state model is developed to obey the principles of fluid dynamics, such as the conservation of mass and energy, as well as Darcy's law governing mass fluxes in a reservoir. Numerical reservoir simulators such as TOUGH2 [4] can be adopted in this and following steps. Steady-state fluid flow due to thermal convection is generally assumed in a reservoir. The natural-state model needs to reproduce the observed static temperature and pressure wireline logging data at the explorational wells while obeying the conceptual model. Calibration of the natural-state model to satisfy these requirements involves adjusting several conditions such as the permeability distributions and boundary conditions and often necessitates numerous trial runs of the reservoir simulator. After completing the natural-state modeling, the transient pressure and temperature responses in the reservoir during well testing are finally simulated to enable history matching and the forecasting of future operational scenarios. These later steps also require trial-and-error simulations and may require going back to earlier steps to reconsider and modify the model.

Planning for the next fiscal year, including decisions with respect to continuing or stopping the explorational and developmental project, as shown in **Figure 1**, requires both updates of the understanding of the geothermal system to determine drilling targets and estimates of the reservoir productivity to evaluate the project profitability. Following the abovementioned standard approach, we can obtain an update of the former at an early step, while an update of the latter becomes available after completion of the final step. As a result, estimates of reservoir productivity are strongly affected by the progress of earlier steps and are often delayed.

2.2 New approach

We can attempt another approach to overcome the difficulties causing delays in the reservoir productivity estimation by advancing inversely in parallel with the standard approach [5]. A comparison between the standard and new approaches is illustrated in **Figure 2**. The new approach first refers to the transient pressure responses during well testing, as well as the other transient responses of the temperature and tracer concentration if possible. In this step, we focus on reproducing these transient responses using a simple reservoir model.

The reservoir model may, at first, be very simple, represented by a single horizontal planar porous medium, as generally assumed in a conventional well test analysis (e.g., [6]). As the study progresses, the reservoir model is extended to become increasingly sophisticated and realistic by considering the dip and strike of the planar porous medium and three-dimensional connections between multiple planar porous media, as common in fracture reservoirs in Japan. The geometry and connectivity of the planar porous media are primarily based on direct observations, such as those made while drilling and logging, as well as pressure interference and tracer testing, rather than referring to geological or other geoscientific interpretations. Once the observed transient responses are successfully reproduced, we can progress to forecasting future operational scenarios and obtain an estimate of the reservoir productivity (i.e., the possible steam and/or brine production rate during an assumed operational period). Appropriately calibrated wellbore models using production logging data are often combined with the reservoir model to forecast future scenarios.

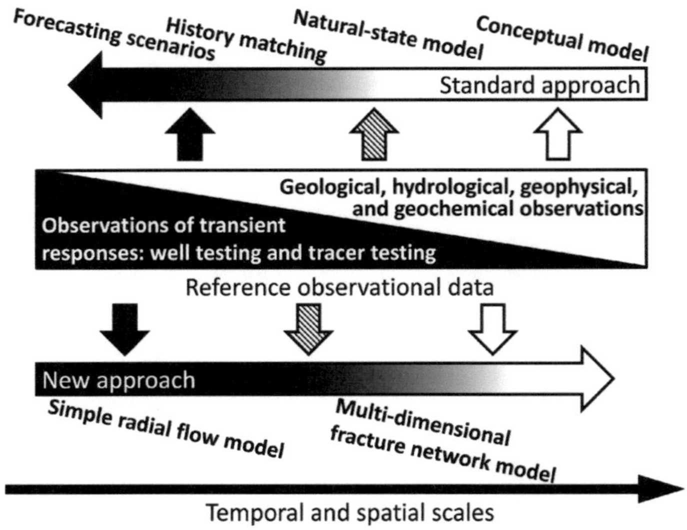

Figure 2.
Comparison between the standard and new approaches.

Accordingly, we can rapidly obtain a first estimate of the reservoir productivity by primarily referring to direct observations, a technique that is free of the conceptual and steady-state models considered in the standard approach. Referring to this estimate, a project manager can prepare a plan for the next fiscal year and make decisions concerning continuing or stopping the project in parallel with the ongoing standard approach. As the understanding of the geothermal system is improved with the standard approach, the reservoir model in the new approach evolves into an ever more sophisticated model that is consistent with the conceptual and natural-state models. The estimate of the reservoir productivity is also repeatedly updated. The estimates and their update history are reported continuously to the project manager; this is advantageous not only for successful project management but also for quantifying the impact of each estimate update. This new approach enables us to improve the efficiency and timeliness of estimating the reservoir productivity and to contribute to on-schedule project management.

3. Implementation

The author developed a mathematical model and numerical code to implement this new approach for a real explorational and developmental project [5]. Instead of a multi-purpose code designed to cover a wide range of conditions, the model and code were designed to be applicable to several specific projects in which the author was involved as a reservoir engineer. Therefore, a type of discrete fracture network model was adopted to represent a single-phase fracture reservoir. As often seen in geothermal prospects in Japan, the fracture network was assumed to be roughly distributed. In other words, at most, a countable number of large fractures or fractured zones with high permeability–thickness products totally or partially intersected a geothermal field (**Figure 3**). Wells in such geothermal fields intersect at most at a few fractures within their drilling depths of approximately 2000 m. Excepting these fractures, formation permeabilities tend to be very low to negligible. Representative examples of such reservoirs can be found in the Takigami [7] and Ogiri [8] fields in southwestern Japan. Each fracture or fractured zone in the model is represented

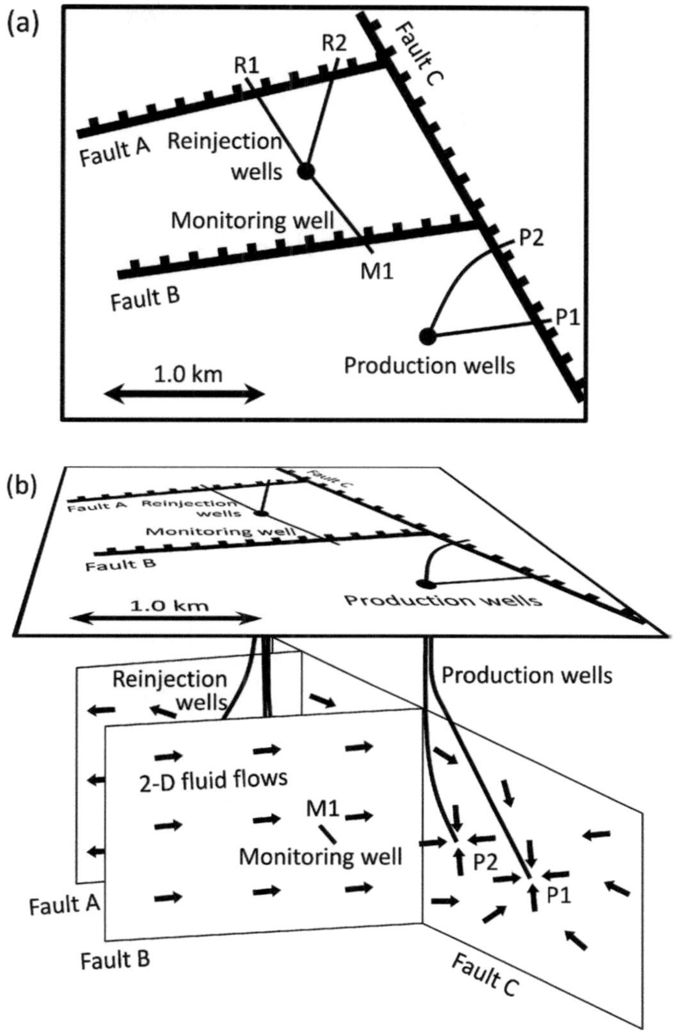

Figure 3.
Conceptual schematic of the reservoir model: (a) an assumed geothermal field and (b) a three-dimensional view of the fracture reservoir model beneath the assumed geothermal field.

by a planar porous medium with a relatively high permeability–thickness product of 10^{-11} m^2 or more. Formations, except fractures, allow only thermal conduction without mass flow.

Let us consider simulating the assumed reservoir illustrated in **Figure 3** as an example. The reservoir consists of three vertical fractures generated by Faults A–C that are represented by planar porous media with the dimensions shown in **Figure 4**. The vertical length of each fracture is assumed to be 2 km based on geological interpretations, while the horizontal length is assumed based on a specific concept of this approach discussed in Section 5. Several production, reinjection, and monitoring wells directionally intersect the fractures. The vertical initial pressure distribution obeys hydrostatic pressure with a specified value of 10 MPa at a depth of 1000 m from the top of the fractures. The initial distribution of the specific enthalpy is uniformly 1085.8 kJ/kg, which indicates an initial reservoir temperature of approximately 250 °C. The top and bottom boundaries of each fracture are modeled with impermeable and adiabatic boundary conditions, while the left boundary of each fracture and the right boundary of Fracture C maintain constant pressure and specific enthalpy values at the initial values.

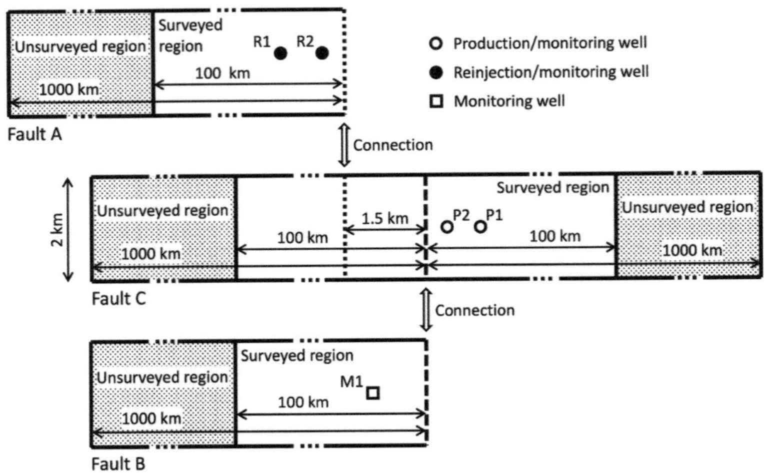

Figure 4.
*Dimensions of the planar porous media representing faults A–C illustrated in **Figure 3**.*

For both the surveyed and unsurveyed regions, the permeability, thickness, and porosity of each planar porous medium representing a fracture are set to 1.0×10^{-11} m^2, 5.0 m, and 20%, respectively, while the thermal conductivity and volumetric heat capacity are set to 3.0 W m^{-1} K^{-1} and 2.0×10^6 J m^{-3} K^{-1}, respectively. We assume that local thermal equilibrium between the fluid in the pores and the rock matrixes within a planar porous medium is reached immediately. This implies that the selection of the thickness value controls the heat exchange efficiency between the fluid and the formation under a constant thickness–porosity product value (i.e., the effective opening width of the fracture). For example, a case with a thickness of 1 m and a porosity of 10% and a case with a thickness of 10 m and a porosity of 1% have the same thickness–porosity product value of 0.1 m; however, the latter case has a larger heat exchange efficiency. This is because the volume of the rock matrix immediately exchanging heat with the fluid in the latter case is approximately 10 times larger than that in the former case. The one-dimensional conductive heat flux in the formation perpendicular to each fracture is also included. The grid size for the numerical simulation in each fracture is a uniform 100 m near the wells and expands exponentially in the horizontal direction.

First, we consider simulating a production test for a month using a production well P1, a reinjection well R1, and monitoring wells M1, P2, and R2. This problem addresses simulating the pressure interference observed at the monitoring wells by referring to the observed flow rates at the production and reinjection wells. We assume that the observed flow rates at P1 and R1 are constant at 250.0 t h^{-1} and 191.6 t h^{-1}, respectively, which implies that the produced reservoir fluid is separated into steam and water under a separator pressure of 0.35 MPaA (**Figure 5a**). The specific enthalpy of the reinjected water is assumed to be 561.5 kJ kg^{-1}, and the reinjected water is composed of saturated water at a pressure of 0.30 MPaA. Then, the pressure interference at each monitoring well is simulated, as shown in **Figure 5b**. In practice, for a real field, we would perform matching of the simulation results with the observations by adjusting parameters such as the permeability, thickness, and porosity of the planar porous media, as well as their network structure. In the cases encountered by the author at several real fields, simulations could accurately reproduce the observed pressure interference after a few tens of trial runs.

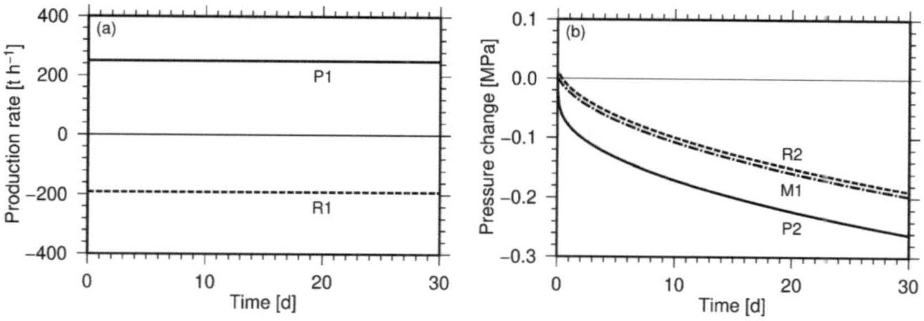

Figure 5.
Conditions and results of the simulation. (a) Assumed flow rates at the production well P1 and the reinjection well R1. Positive values indicate production, while negative values indicate reinjection. (b) Simulated pressure interference at the monitoring wells M1, P2, and R2.

4. Connecting wellbore flow and reservoir models

Next, a wellbore flow model is connected to the reservoir model; this is necessary to simulate and forecast future operational scenarios under specific conditions, such as a constant wellhead pressure. The author's numerical code implements this connection via two procedures. One is based on tabular data of the production rate depending on the feed zone pressure and the specific enthalpy. The other applies a highly refined local grid to simulate steep changes in the reservoir pressure around the wellbores.

Let us extend the reservoir model described in Section 3 by connecting it to a wellbore flow model. Using an appropriate wellbore flow simulation code (e.g., [9–14]), we assume that the production rate at a production well P1 with a constant wellhead pressure depends on the feed zone pressure and the specific enthalpy, as shown in **Figure 6**. Referring to the discretized tabular data, the code dynamically determines the production rate corresponding to arbitrary values of the pressure and the specific enthalpy via a bicubic interpolation. The reinjection rate at the reinjection well R1 is also dynamically determined by referring to the production rate at P1. Note that the code can only assume the steady-state wellbore flow represented by the tabular data. Simulating unstable transient wellbore flows, which is often a problem in operational power plants, connected to a reservoir model is a goal for future studies.

The pressure distribution covering the overall reservoir, including in the vicinities of wellbores, is simulated seamlessly using the highly refined local grids described in detail by [15]. The local grid defined around a wellbore enables steep pressure changes generated by production and reinjection at the well to be simulated by adopting grid

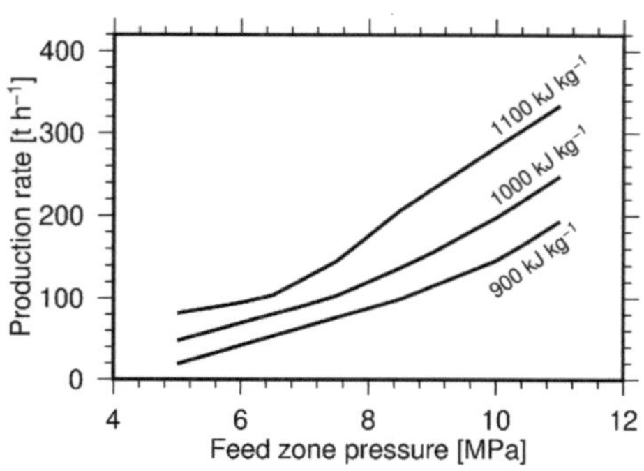

Figure 6.
Assumed production rate for the production well P1 depending on the feed zone pressure and the specific enthalpy.

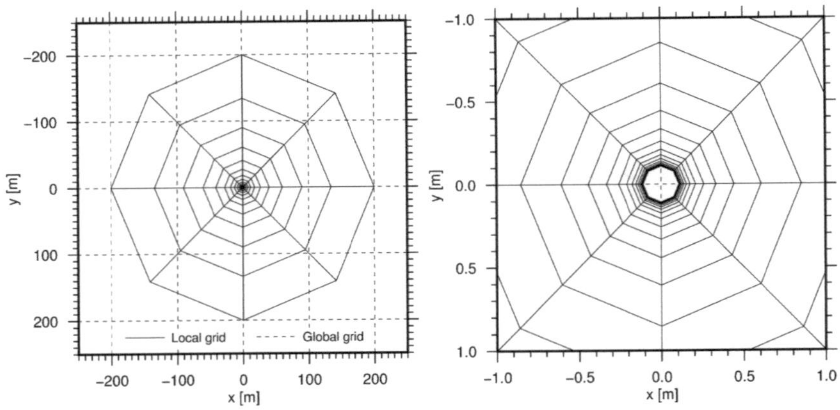

Figure 7.
Highly refined local grid around an 8.5-inch wellbore superimposed on a global grid shown on different scales.

sizes down to 1 mm, which is several orders of magnitude smaller than the size of the global grid covering the total planar porous medium. As shown in **Figure 7**, the ring-shaped local grid has external and internal boundaries. The variable values at the external boundary are dynamically determined by the values distributed in the global grid and are interpolated using a bilinear interpolation. Conversely, the variable values at the inner boundary corresponding to the wellbore surface are dynamically determined by the mass and enthalpy flow rates between the wellbore and the reservoir. When considering the skin effect, an extra pressure loss is considered between the internal boundary and the inside of the wellbore. When assuming the dependence of the production rate, as described in the previous paragraph, the production rate is determined as a solution of a coupled problem between this dependence and the fluid flow in the local grid. The determined production rate is referenced by the global grid to simulate the fluid flow in the global grid. Therefore, simulations in the global and local grids are dynamically coupled by referring to each other.

The simulated production and reinjection rates, as well as the pressure interference, using the extended model are shown in **Figure 8**. In this model, the production well P1 is equipped with tabular data for the production rate and a highly refined local grid assuming a hole size of 8.5 in. Selecting a value of 1.0 for the skin factor of P1, the simulated production and reinjection rates are similar to those assumed in the model described in Section 3. In fact, the production rate decreases gradually from 270.6 t h^{-1} to 243.0 t h^{-1} over the simulation period of 30 d. In this step of practical modeling connecting the wellbore and reservoir models, only the skin factors are

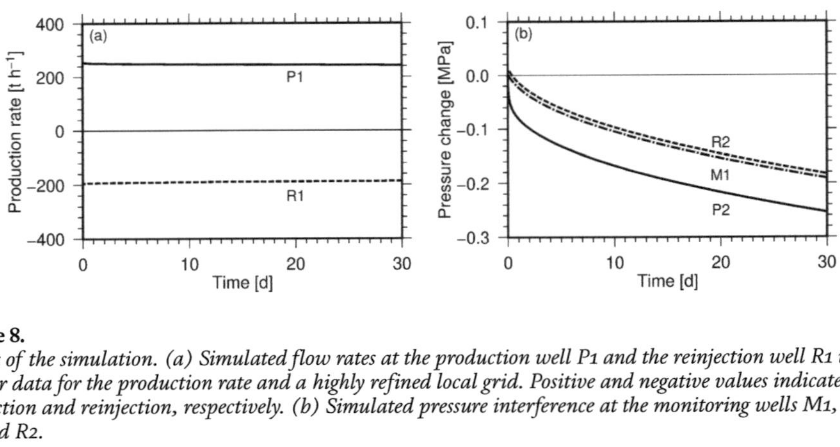

Figure 8.
Results of the simulation. (a) Simulated flow rates at the production well P1 and the reinjection well R1 using tabular data for the production rate and a highly refined local grid. Positive and negative values indicate production and reinjection, respectively. (b) Simulated pressure interference at the monitoring wells M1, P2, and R2.

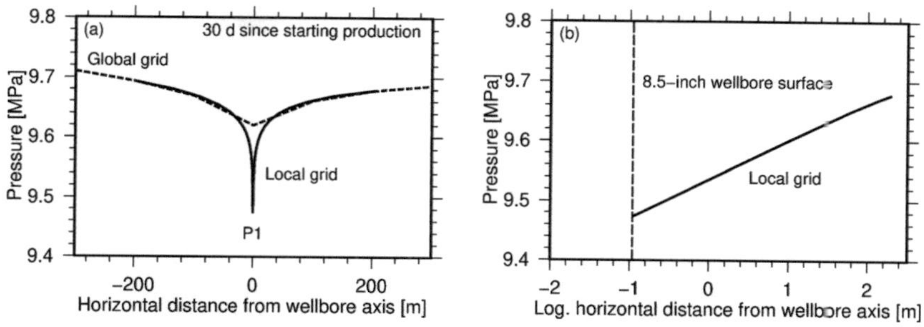

Figure 9.
Horizontal pressure distribution at the feed zone depth of the production well P1. (a) Pressure distribution on a linear scale simulated using the local grid (solid line) superimposed on that simulated using the global grid (broken line). (b) Pressure distribution on a logarithmic scale simulated using the local grid.

modified to match the simulated production rates with the observations, and the other parameter values of the planar porous media, such as permeability, thickness, and porosity, are maintained. In other words, the procedures in Sections 3 and 4 are straightforward.

The simulated pressure distribution around the production well P1 is shown in **Figure 9** on linear and logarithmic scales. The pressure distribution simulated using the local grid is smoothly connected to that simulated using the global grid and successfully reproduces a steep pressure drop in the vicinity of P1 (**Figure 9a**). On the logarithmic scale, it can be seen that the pressure increases proportionally to the logarithmic distance from the wellbore axis (**Figure 9b**), which is consistent with the solution of the problem assuming radial flow from a line-source adopted in the conventional well test analysis (e.g., [6]).

5. Estimating reservoir productivity

Finally, let us estimate the productivity of the above-discussed reservoir model by forecasting operational scenarios. We consider 15-year scenarios using the production wells P1 and P2, reinjection wells R1 and R2, and monitoring well M1. Both production wells obey the wellbore flow model described in the previous section. Forecasting scenarios for 15 years based on production tests for, at most, a few months involves uncertainty. We attempt to quantify this uncertainty by defining the surveyed and unsurveyed regions as illustrated in **Figure 4**.

Performing a longer production test, transient reservoir pressure responses are constrained by the wider spatial range reservoir properties, as discussed using the radius of investigation in the conventional well test analysis. We define the surveyed region as the region that constrains the simulated pressure responses when matching with observations, while the unsurveyed region is too distant to constrain the simulated pressure responses. Examining the effects of modifying the reservoir properties on the simulated pressure responses, the boundaries between the surveyed and unsurveyed regions are determined by trial and error. The boundaries move farther when performing longer production tests, indicating that the uncertainty in the forecasting scenarios becomes smaller. Note that defining the surveyed and unsurveyed regions, as well as extending the reservoir model to a huge horizontal distance, are symbolic parameterization methods for the uncertainty in terms of the transient pressure responses and do not include geological or other geoscientific insights.

Once we define the unsurveyed region, natural recharge and/or discharge over the boundaries between the surveyed and unsurveyed regions can be quantified

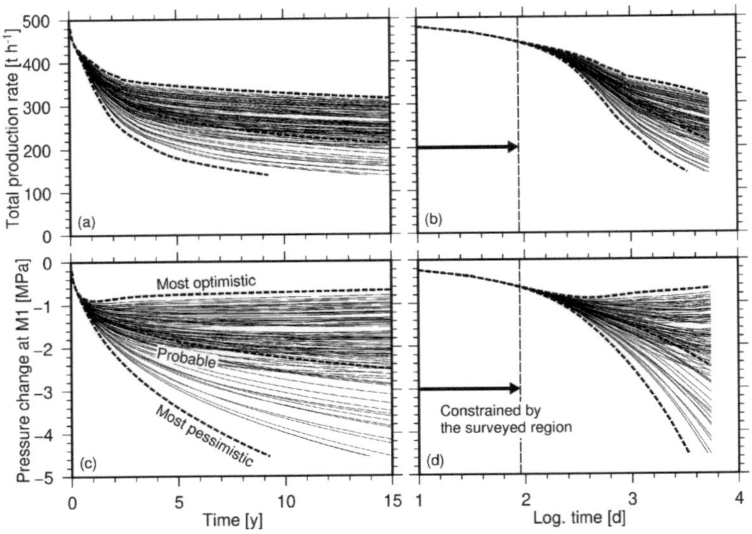

Figure 10.
Temporal changes in the total production rates at the production wells P1 and P2, as well as the pressure changes at the monitoring well M1, for 100 trial runs. Because the code is only capable of simulating single-phase reservoirs, runs whose reservoir pressure drops below the boiling pressure are terminated prior to 15 y.

using the permeability in the unsurveyed region as a symbolic parameter. Theoretically, the most optimistic case is to take the limit as the permeability approaches infinity, which is equivalent to assuming constant-pressure boundaries between the surveyed and unsurveyed regions. Conversely, the most pessimistic case is to set the permeability to zero, indicating impermeable boundaries. Modifying the permeability between zero and infinity, we can continuously control the magnitude of natural recharge and/or discharge. Even though, in the strictest sense, we have no information about the unsurveyed region, a probable case can be defined by giving the unsurveyed region the same permeability value as that in the surveyed region. All the parameters in the unsurveyed region, except the permeability, are assumed to be equal to those in the surveyed region.

In this chapter, we assume that the boundaries between the surveyed and unsurveyed regions are approximately 100 km from the wells (**Figure 4**). In the unsurveyed regions, the permeability values are randomly and independently selected in a range from 1.0×10^{-13} m^2 to 1.0×10^{-9} m^2 at each planar porous medium, while the permeability in the surveyed region is constant at 1.0×10^{-11} m^2. The probability distribution of the selection is assumed to be uniform in this range on a logarithmic scale. Under these conditions, temporal changes in the simulated total production rates from the two production wells P1 and P2, as well as the pressure changes at the monitoring well M1, for 100 trial runs are shown in **Figure 10**. For reference, the probable, most optimistic, and most pessimistic cases are also shown; these cases assume uniform permeability values in the unsurveyed regions of 1.0×10^{-11} m^2, 1.0×10^{-9} m^2, and 1.0×10^{-13} m^2, respectively. The simulated changes in the production rates and the reservoir pressures for all trial runs exhibit unique changes for approximately three months following the start of production. This indicates that the surveyed region extending 100 km from the wells constrains the changes during this period.

Summarizing the results of all the trial runs, we obtain monomodal frequency distributions for the production rates and the pressure changes, as shown in **Figure 11**. The medians of these distributions indicate a slightly more optimistic case (i.e., a larger production rate and a smaller pressure interference) than the probable case. This implies that a relatively high permeability value occurring in

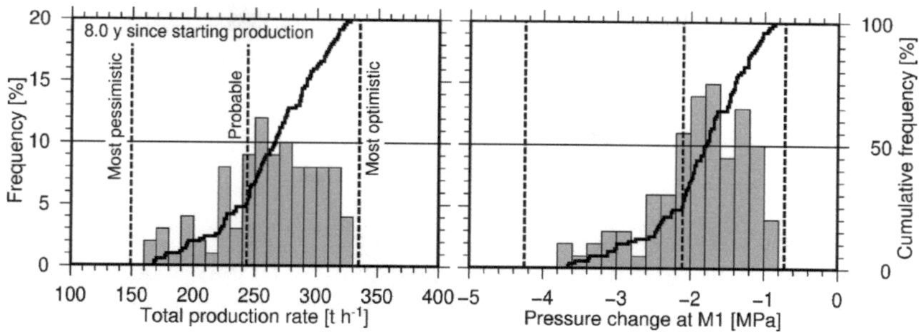

Figure 11.
Frequency distributions for the total production rates at the production wells P1 and P2, as well as the pressure changes at the monitoring well M1, after production for 8.0 y for 100 trial runs.

one unsurveyed region can result in a large overall reservoir productivity regardless of the low permeabilities in the other unsurveyed regions.

Accordingly, we obtain a present estimate of the reservoir productivity. Accounting for the power generation capacities corresponding to the forecasted production rates, the price of electricity, and the costs for the entire project, we can calculate the series of cash flows during the project period, as well as several profitability indices, such as the net present value and the internal rate of return. If the reservoir model is updated as a result of the in-parallel progress of the standard approach, the frequency distributions shown in **Figure 11** are also immediately updated via the procedures described in this chapter. The author mentions again and emphasizes that the new approach proposed in this chapter cannot stand alone because the standard approach, which exhaustively studies the geothermal system in terms of several geoscientific fields, is also essential. The combination of these approaches enables us to make the most of the limited time available during active explorational and developmental projects.

6. Forecasting temperature changes

Here, we briefly remark on forecasting temperature changes. As mentioned above, this chapter's approach aims to directly adhere to the observed transient responses as much as possible rather than referring to geological or other geoscientific interpretations. However, it is generally difficult to effectively forecast temperature changes in a reservoir over the decades of an assumed operational period. This is because promising prospects often do not exhibit detectable temperature changes at both production and monitoring wells during production tests. Even though tracer testing provides useful insights into advection from reinjection to production wells, it is not always sufficient to constrain the heat exchange efficiency between the fluid and the formation (i.e., it is not sufficient to determine the thickness of a planar porous medium, as mentioned in Section 3) while flowing in a reservoir.

To overcome this limitation, we need to develop measures to determine the heat exchange efficiency between the fluid and the formation rather than improve the modeling techniques. One possible technique may be to use dual tracers with different thermal resistivities, as proposed by [16]. The authors of that study proposed that temporal changes in the concentration ratio of a mixture of thermo-resistant and thermo-sensitive tracers depend on the temperature. Using this principle, we can monitor the temperatures experienced by reinjected fluid while it flows between the reinjection and production wells; this depends on the heat exchange efficiency between the reinjected fluid and the formation. A simulation from this viewpoint using dual tracers by [5] demonstrated the detection of two flow paths with different temperatures in a reservoir. Another possible technique may be to perform a push-pull test, which compares the temperatures of injected and pumped-up fluids using a single well.

7. Conclusions

The author has developed an approach to estimate reservoir productivity under severe schedule limitations. Such limitations can originate from limited access to sites for drilling and well testing under snowy and mountainous conditions. To make the most of the limited time available, the new approach progresses in parallel with the conventional standard approach with an inverse approach using the transient responses observed while performing production tests, which are referenced in the final step of the standard approach. Combining these approaches, estimates of reservoir productivity can be rapidly generated. This feature is of value to successfully manage explorational and developmental projects.

Assuming a practical geothermal field, the author has demonstrated the procedures of this new approach. The procedures are straightforward: the pressure interference is simulated at the monitoring wells, the production and reinjection rates are simulated by combining the wellbore and reservoir models, then future operational scenarios are forecasted. By defining surveyed and unsurveyed regions, the reservoir model strictly divides the simulation period into an earlier period constrained by observations through the parameters in the surveyed region and a later period with no constraints. Performing a number of trial runs while randomly selecting permeability values in the unsurveyed regions, we can obtain frequency distributions for estimates of the reservoir productivity and successfully make project management decisions.

Acknowledgements

We thank Martha Evonuk, PhD, from Edanz Group (https://en-author-services.edanz.com/ac) for editing a draft of this manuscript.

Author details

Mitsuo Matsumoto
Department of Earth Resources Engineering, Faculty of Engineering, Kyushu University, Fukuoka, Japan

*Address all correspondence to: matsumoto@mine.kyushu-u.ac.jp

References

[1] International Energy Agency Geothermal. 2019 IEA Geothermal Annual Report. Taupo: International Energy Agency Geothermal; 2020. 168 p.

[2] Grant, M.A., Bixley, P.F. Geothermal Reservoir Engineering. 2nd ed. Burlington: Academic Press; 2011. 378 p. DOI: 10.1016/C2010-0-64792-4

[3] DiPippo, R. Geothermal Power Plants: Principles, Applications, Case Studies and Environmental Impact. 2nd ed. Oxford: Butterworth-Heinemann; 2008. 520 p. DOI: 10.1016/B978-0-7506-8620-4.X5001-1

[4] Pruess, K., Oldenburg, C., Moridis, G. TOUGH2 user's guide, version 2. Berkeley: Lawrence Berkeley National Laboratory; 1999. 197 p. LBNL-43134. DOI: 10.2172/751729

[5] Matsumoto, M. A single-phase reservoir simulation method based on a roughly distributed and highly permeable fracture network model with applications to production and reinjection problems. Geothermics. 2020; 84: 101744. DOI: 10.1016/j.geothermics.2019.101744

[6] Dake, L.P. Fundamentals of Reservoir Engineering. Developments in Petroleum Science 8. Amsterdam: Elsevier; 1983. 462 p.

[7] Furuya, S., Aoki, M., Gotoh, H., Takenaka, T. Takigami geothermal system, northeastern Kyushu, Japan. Geothermics. 2000; 29: 191-211. DOI: 10.1016/S0375-6505(99)00059-0

[8] Goko, K. Structure and hydrology of the Ogiri field, West Kirishima geothermal area, Kyushu, Japan. Geothermics. 2000; 29: 127-149. DOI: 10.1016/S0375-6505(99)00055-3

[9] Miller, C.W. WELBORE user's manual. Berkeley: Lawrence Berkeley Laboratory; 1980. 48 p. LBL-10910. DOI: 10.2172/6334005

[10] Pritchett, J.W. WELBORE: a computer program for calculating flow in a producing geothermal well. La Jolla: S-Cubed; 1985. Report SSS-R-85-7283.

[11] Garg, S.K., Pritchett, J.W., Alexander, J.H. A new liquid hold-up correlation for geothermal wells. Geothermics. 2004; 33: 795-817. DOI: 10.1016/j.geothermics.2004.07.002

[12] Aunzo, Z.P., Bjornsson, G., Bodvarsson, G.S. Wellbore models GWELL, GWNACL, and HOLA. Berkeley: Lawrence Berkeley Laboratory; 1991. 102 p. LBL-31428. DOI: 10.2172/937440

[13] Pan, L., Oldenburg, C.M. T2Well—an integrated wellbore–reservoir simulator. Comput. Geosci. 2014; 65: 46-55. DOI: 10.1016/j.cageo.2013.06.005

[14] Matsumoto, M., Itoi, R., Fujimitsu, Y. Theoretical study of conditions for generation of periodic wellbore flow due to inflow of a lower-enthalpy fluid. Geothermics. 2021; 89: 101948. DOI: 10.1016/j.geothermics.2020.101948

[15] Matsumoto, M. Connecting wellbore and reservoir simulation models seamlessly using a highly refined grid. In: Proceedings of the 43rd Workshop on Geothermal Reservoir Engineering; 12-14 February 2018; Stanford. Stanford: Stanford University; 2018.

[16] Adams, M.C., Davis, J. Kinetics of fluorescein decay and its application as a geothermal tracer. Geothermics. 1991; 20: 53-66. DOI: 10.1016/0375-6505(91)90005-G

Chapter 6

The Hot Springs of Central Northern Algeria Hydro Geochemical and Therapeutic Aspects: Direct Applications and Therapeutic Value

Mébrouk Benziada

Abstract

This Science article summarizes the preliminary work carried out by the Renewable Energy Development Centre under the National Research Project in the Central North of Algeria from 2013 to 2016 to explore for geothermal resources and hydrogeology and hydrogeochemical and therapeutic aspects. The geology is very complex in this region and it determines the thermal water reservoirs. The value and importance of the thermal springs in Algeria is very significant. This study will be addressed by means of conventional chemical analyzes of the main hot springs major elements in the north central region of Algeria. Hydrogeochemical prospecting was carried out in detail is briefly exposed and the main results are described in particular regarding the hot springs of the Centre North of Algeria. The existing geothermal potential in Algeria is operated primarily for the balneotherapy and some applications for aquaculture. In this study, we will apply the hydrogeochemical techniques to the hot springs of the Centre North of Algeria. To promote this energy source which will certainly have a socio-economic interest, it is important to know the geothermal gradient in this region and the hydrogeological and physico - chemical characteristics of the main hot springs particularly temperature, flow rate and the elements major chemical. The thermal springs of north central Algeria belong to the Tellian domain, characterized by a complex geological structure and active tectonics. The thermal spring of Hammam Bouira is strongly mineralized, it presents a sodium chloride chemical facies. These facies are generally linked to Triassic evaporites. The thermal waters of the study area are characterized by a hyper mineralized facies (4 to 20 mS/cm at 25° C), sodium chloride type source Hammam Melouane and source Hammam Bouira. The mineralization of these thermal springs is acquired from contact with the evaporite formations of the Triassic, mainly by the dissolution of halite and gypsum. Known for its geological, structural and tectonic complexity, the northwestern region of Algeria has very important thermal manifestations. The presence of thermal springs in this region are linked to the existence of faults. Ten sources of spas were analyzed in the laboratory of the National Tourist Company in 1984. The chemical analysis of the major element concentrations (Ca2 +, Mg2 +, Na +, K +, Cl-, SO4 2-, HCO3 - and NO3 -) and the physico-chemical variables (temperature, electrical conductivity, dry residue and pH). The physico-chemical composition of warm waters in this region shows a very varied chemical facies due to the complexity of geology. The geothermal applications are very diverse: - The balneology, aquaculture, heating greenhouses, air conditioning habitat and production of electricity. The use of geothermal energy helps to preserve natural resources while reducing CO2 emissions related to energy production and to cure certain diseases. Direct use of geothermal energy in agriculture, spas and domestic heating is possible.

Keywords: Hydrogeology, Hydrogeochemistry, Geothermal gradient, Therapeutic aspects, geothermal potential, low-energy, direct applications, health

1. Introduction

Algeria belongs to the Mediterranean basin; the Sahara occupies the major part of the country.

Algeria is located between Morocco and Tunisia forming the Maghreb (**Figure 1**).

This scientific article shows an overview of the geology, geothermal resources; chemical analyzes of water thermals therapeutic applications in Algeria.

Figure 1.
Location map of Algeria in the world.

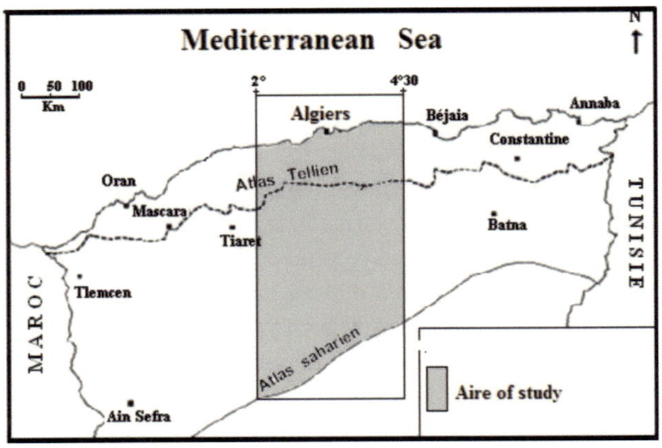

Figure 2.
Geographical location of the study area.

In Algeria, balneotherapy is practically the only direct use of geothermal energy. Yet the country has more than two hundred hot springs distributed throughout the northern region of Algeria. About one-third (33%) have temperatures above 45° C. There are even sources at high temperatures of the order of 98° C in Guelma. Algeria is determined to diversify its economy, to free itself from its heavy dependence on oil revenues, by turning to the development of renewable energies. The study area is located north central Algeria between longitudes 2° and 5°. It is limited to the north by the Mediterranean Sea, the South by flexure south Atlas The geology of hydrothermal sites Hammam Melouane, Hammam Righa, Hammam Médéa and Hammam Bouira (a) (b) and is described in our field trips to determine the hydrogeological and hydrogeochemical characteristics of the underground environment (**Figure 2**).

2. Geology of the north of Algeria centre

Region consists of structural and sedimentary units showing an imprint of the Alpine tectonics. From north to south there are: the Tell Atlas, varied and complex area. It includes an internal zone and an outer zone formed of non-native land (thrust sheets). Between the two Atlas (Tellian and Sahara) are flush with the High Plains ending to the east by the chain of Hodna and continue to the west by the Oran Meseta (**Figure 3**). The sedimentary formations covers are of Mesozoic and Cenozoic age and based on a diverse base involved in the fold [1].

The warm and sulfurous waters of Hammam Ksana emerge from the bowels of the geological layers of the Triassic and Jurassic of the rocky peaks of the Bibans. The source it self is located at 500 m altitude on a slope of Wadi Tazdart, douar of El Mehir.

Tellian area:
Characterized by a stack of thrust sheets with associated intra-mountain basins Sedimentary formations blankets are Mesozoic and Cenozoic age and based on a diverse base involved in the folding.
Saharan area:
Relatively stable or tectonics is less pronounced (**Figure 4**).

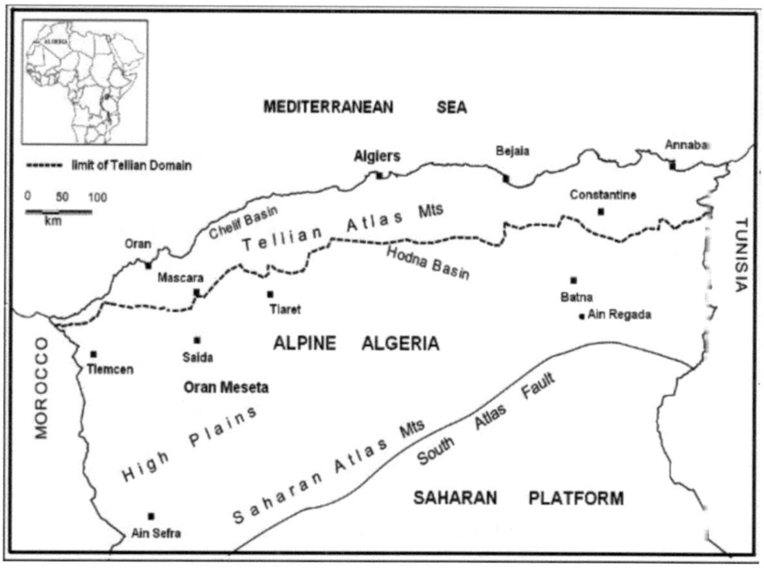

Figure 3.
Geological units of the northern Algeria.

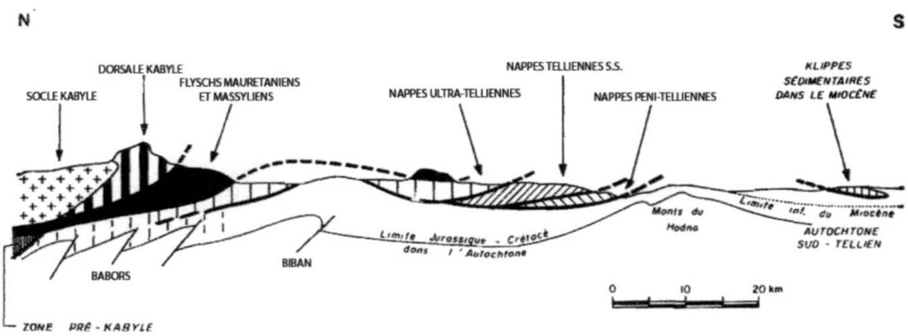

Figure 4.
Relationship between the different structural units of the chain of Maghrebides (modified from Durand-Delga, 1969 Bibans and Babors who were considered the aboriginal also include tablecloths Tellian).

2.1 Internal domain

Kabyle massive base cristallophylliens metamorphic (gneiss, marble, amphibolite schists and mécaschistes) [2]. This stand exposed in the massive Chenoua (Tipaza) and Great Kabylia (**Figure 3**).

2.2 Domain flyshs

Consisting of layers of Cretaceous-Paleogene flyshs.

2.3 External domain

Tellian domain consists of a set of non-native groundwater; characterized by a marly facies of Middle Cretaceous age has Neogene.

3. Methodology and materials

3.1 Work in the office

Make an inventory of oil drilling reconnaissance work. The counting of polls reporting was used to select thirty boreholes with their BHT (Bottom Hole Temperature) measured at different depth logging operations to establish a geothermal gradient map. A geological and hydrogeological survey of the thermal springs of writing is based on previous geological research studies.

3.2 Fieldwork

In our scientific research work in the field, we made direct measurements of the Ph, conductivity and temperature of the thermal water and take samples of the water samples of the hot springs of four sites (H. Melouane, H. Righa, H.Médéa and H.Ksénna) to analyze in the laboratory of the National Agency of Water Resources in Algiers to determine their physicochemical characteristics. The results are shown in **Table 1**. We selected four most important thermal springs which were the subject of several measurement campaigns (2014–2015). Work performed and results obtained.

DOI: http://dx.doi.org/10.5772/intechopen.97803

	pH	Cond µmho/cm	Rs g/l	T°C	Flow l/s	Facies Chemical
H. Melouan	6.4	20000		40	—	Chloride- sodium
H.Righa	7.53	4000	3.4	64	20	Sulfate- calcium
H.Médéa	7	2800	1.4	38	—	Bicarbonat-sodium
H.Bouira (a)	6.5	17200	5.9	63	20	Chloride- sodium
H.Bouira (b)	—	—	—	80	1.2	Chloride- sodium

Table 1.
Physical and chemical characteristics of the thermal source in the study region (2014–2015).

4. Results and discussions

4.1 Map preliminary geothermal gradient Northern Algeria centre

The geothermal gradient map sets from the temperature data stored in oil drilling during logging operations. It shows a geothermal anomaly of about 4° C/100 m in the Djelfa region (**Figure 5**). It serves as a support base for future projects in the applications of geothermal energy [3].

4.2 Hydrogeochemistry of hot springs

The study area is characterized by the emergence of numerous hot springs area linked to major geological accident (faults) [4–6]. The temperature of these sources usually varies from 20 to 64° C. The chemical composition dominant is sodium chloride-sulfated.

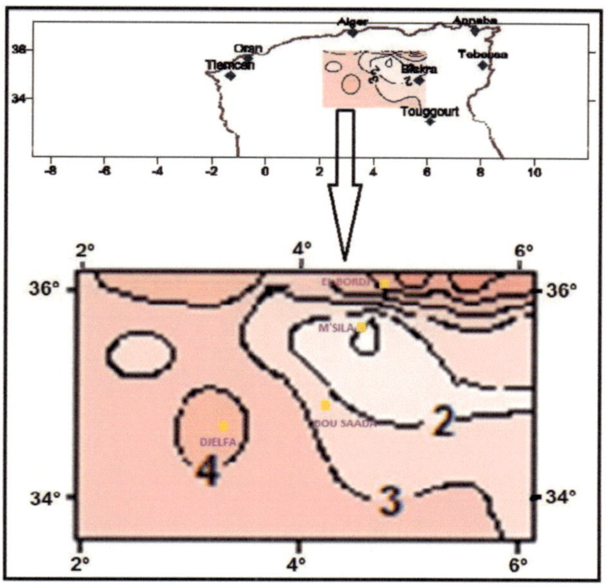

Figure 5.
Map preliminary geothermal gradient northern Algeria Centre.

4.3 Results hydrogeochemical hot springs centre Northen of Algeria (2013–2015)

The Jurassic limestones of the Algerian North, which constitute important geothermal.

reservoirs, give rise to more than five hot springs located mostly in the north-east and northwest regions of the country [7]. These sources have temperatures above 40° C. That of Hammam Bouira (b) (80° C) is the most important. These natural emergences are generally the losses of existing reservoirs, with a flow rate of the order of 2 m^3/s. The temperature of the hot springs in the study area varies from 38° C to 80° C which corresponds to a low energy enthalpy. The pH of the thermal waters ranges from 6.4 to 7.53 and generally neutral rate is between 3 and 20 l/s (**Table 1**). The conductivity of the largest thermal water is the source of Hammam-Melouane with a value of 20000 μmhoS/cm and the lowest is that of Hammam-Righa with 4000 μmhoS/cm. The highest-dry residue is that of Hammam-Bouira-b source with 5.8 g/l.Sources H.Melouane and H; Bouira-b have the same chemical profile of a chloride-sodium kind while that the source of H.Righa has a sulfated lime facies and source H.Médéa a sodium bicarbonate-facies (**Figures 6** and 7).

4.4 Hydrogeological hot springs central northern of Algeria

4.4.1 Hammam-Melouane

Griffins are aligned along a major geological accident oriented E-W, by contacting the Cretaceous and Lower.

4.4.2 Hammam-Righa

The hydrogeological study reveals the existence of two aquifers, the most important is represented in Zaccar Chergui by cracked limestone of the Upper Jurassic, strongly mineralized and intensely karstified. These limestones are based on the schist-quartzite waterproof primary series and are an important reservoir whose static level is close to the coast 700 m.

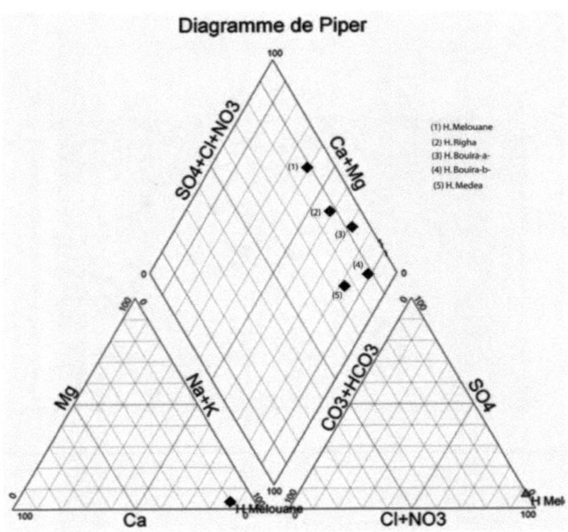

Figure 6.
Classification of the thermal springs (piper diagram).

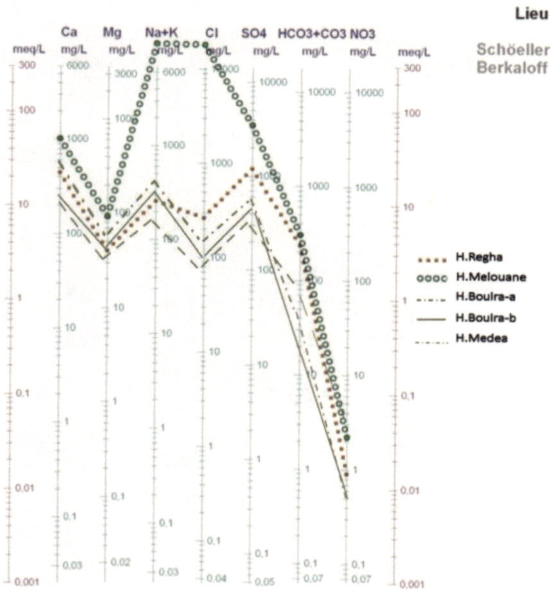

Figure 7.
Classification of the thermal springs (diagram Schoeller-Berkaloff).

4.4.3 Hammam-Medea

A thermal spring with a flow rate of 3 liters/second. The waters of Hammam Medea emerge the feet of the northern flank of Jebel El Gharbi Souebah They are very rich in mineral and weakly conductive. The most dominant ion correspond bicarbonate and sodium. The mineralization of these waters is acquired by the leaching of geological formations during their ascent. The Cretaceous formations, limestone of the Turonian and Vraconian are the only aquifer can supply the hot spring.

4.4.4 Hammam-Bouira (a) (b)

The waters of Hammam Ksana (Bouira) emerging south of the plain of El-Asnam. They are loaded with minerals and highly conductive. The most dominant ions are chloride and sodium. The mineralization of these waters is acquired by alteration lithologies traversed during circulations of underground fluids. The aquifer can match the Cenomanian limestone.

5. Geothermal resource in Algeria

Geothermal resources are various low temperatures. They are located in northern Algeria and northern Sahara [8].

In the north, the reservoirs are complex and discontinuous. They consist of facies (limestone, sandy limestone and sandstone) of the Mesozoic and south, a continuous tank is composed mainly of sandstone [9].

He was appointed tableclot Albian [10].

The distribution of thermo-mineral waters, in Algeria, is very irregular. These sources are increasing in number gradually as one approaches from the east. This distribution seems to follow that of the ore deposits. Thus it has about twenty mineral springs in the region of Oran, about 40 in the Algiers and 150 in the region of Constantine.

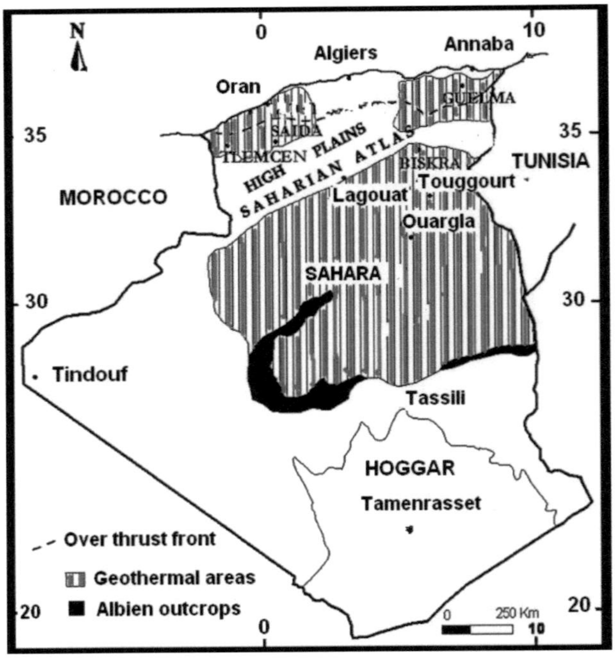

Figure 8.
Main geothermal area [11].

In the North, in part including the Saharan Atlas to the sea, emerging more than 200 hot springs. The water temperature varies between 22 and 98° C and mineralization total dissolved salt varies between 4 and 10 g/l. The reservoirs are typically at depths between 1500 and 2500 m.

South, the water Albian covers an area of 600,000 km^2. It is semi-free in the western and captive and warm only in the eastern part (**Figure 5**) [11]. The roof of the aquifer to the east is located between 1000 m and 2600 m, the average temperature of water is 60° C and an operating rate of 4 m^3/s, the water Salinity is 3 g/approx.

The geothermal potential to dewatering, by summing the total flow of the hot springs flow from operating Albian aquifer is over 700 MW (**Figure 8**).

6. Use of geothermal energy

6.1 Agriculture and geothermal energy

The agricultural field is compatible with the use of geothermal energy [12]. The heating of greenhouses, for example vegetable farms, can be carried out by heat from the ground and deep boreholes to collect geothermal hot water. The two main applications of geothermal energy in agriculture are fish farming and agricultural greenhouses. For fish farming, an increase in temperature by a few degrees and above all its maintenance at a constant level produces an increase in metabolism in fish and crustaceans.

6.2 Geothermal energy for health

Geothermal energy is good for health! As a natural energy source, geothermal energy is not harmful to humans and their environment, unlike some fossil fuels [13]. Geothermal water is also known for its benefits and is particularly used in spas to treat rheumatism. Hydrotherapy was one of the first applications of geothermal energy during antiquity, recognized for its health qualities. Today, it remains an excellent natural remedy for thermal cures.

7. Spas in Algeria and their healing powers

Over 80 spas are operated across the country, including five major national. 200 thermal springs inventoried across the country. Each source including therapeutic advantages. Spas most requested by the Algerian population for various therapeutic treatments (rheumatism, dermatology, gynecology etc. ...). Thermal establishments in Algeria have SPE and modern facilities with highly qualified medical and para-medical to the contribution of patient care and hydrotherapy. Thermal medicine offers an original approach health is not based only on the treatment of symptoms. The major advantage of the spa medicine is based on the overall pathology support, providing prevention as an important place that relief. Some treatments, such as jetted tubs will act as a leg massage to stimulate blood circulation, more oxygen to the muscles and reduce pain experienced. Due to the quality of its waters, for several decades there has been a craze for spa tourism in Algeria. Depending on the conditions being treated, customers opt for a stay in spas waters renowned for their healing qualities.

7.1 Therapeutic hot springs North Centre of Algeria

7.1.1 Hammam-Melouane (Blida)

Located 37 km south of Algiers at an altitude of 105 m - Nature and Thermality water: ferrugineous and chloride, with sodium, very little limestone, 29° to 41°C. Indications: Rheumatology, Dermatology.

7.1.2 Hammam Righa (Ain Defla)

Located 100 km south-west of Algiers, at 520 m altitude. It is located in a green and forested region appreciated for its unpredictable weather [14, 15]. Thermal water rich in minerals, Salinas, sulfated, calcium 39° to 67°C. Spa hammam Righa is a mecca for full health, adapted to different therapeutic orientations. Some sources have disappeared during the earthquake of 1980. Therapeutic Indications: arthritis, rheumatic diseases, nervous disorders, injuries and trauma, anemia, hepato-renal insufficiency, chlorosis, embarrassment digestive functions, in general all diseases know because the loss of blood. The curing techniques: general and local bathing, jet showers and ablutions, underwater massages. Dry massage, electrotherapy, rehabilitation and paraffin wraps.

7.1.3 Hammam-Médéa

He warm and therapeutic waters are excellent and strongly recommended for the treatment of gynecological, epidermal, neurological diseases as well as rheumatism.

7.1.4 Hammam Bouira (a) (b)

A thermal water with its important hydrothermal emergence with proven therapeutic virtues.

Rheumatism, neuropathy, myopathy, and certain affections ENT, intestinal, dermatosis, sequel of burns, arteriopathy, varicose veins, asthma, paralysis, chronic bronchitis. Certains pathologies are treated by the thermal water of Hammam Bouira (a): herpes, mycoses, rheumatism, low back pain, certain arthroses, affections uro-genitales etc. ...

7.2 Habitat heating

In the case of a district heating network of large, deep aquifer Albian can be exploited by a geothermal doublet. Depending on the type of installation, geothermal covers 50–100% of the heat demand. After using the hot springs for heating, reclaimed water can ensure the supply of thermal pools complex Zelfana which falls within the field of health and recreation.

7.3 Agriculture

7.3.1 Aquaculture is a growth area in Algeria

The two main applications of geothermal energy in agriculture are the fish and greenhouses [16]. Greenhouse crops are an attractive option because the energy requirements are high.

As regards fish farming, an increase in temperature of a few degrees above and keeping it at a constant level produces an increased metabolism of fish and crustaceans. In the field of food manufacturing, temperatures between 40° C and 100°C is used for dehydrating fruits and vegetables. From 60° C, the heated air can be used for drying agricultural products, fish and timber.

7.4 Industry

The frost of large industrial buildings can be achieved by a moderate temperature geothermal resource, but most hot water needs or steam industry is between 100° C and 200° C. If the geothermal resource is less than 100° C, it will be used to preheat iron water, whose temperature will be increased by means of a heat pump. Many processes require large amounts of hot water, such as pulp and paper, textile washing, extraction of chemicals or evaporation of concentrated solutions [17].

Thus it is easy to see, Algeria contains an abundance of thermo-mineral sources of various compositions and meet all the needs of modern therapeutics. It is highly probable that most of these radioactive sources are: education, barely sketched, can believe it. It is therefore obvious that most inhabitants of Algeria could find, there, in the colony, mineral waters they need [18].

8. Conclusions

The center-north of Algeria is characterized by an abundance of springs and by high heat flow values [19]. Hot waters, seem essentially controlled by tectonics, because most thermals springs are located on faults or abnormal contacts. The heat flow study shows one important anomaly:

An anomaly situated in the region of Djelfa. The geophysical study highlights the Djelfa region is most favorable to the future exploitation of geothermal energy with a gradient of 4° C/100 m. The numerous thermal springs northern central Algeria, where better exploitation of these natural resources is an opportunity for a good investment in the housing niche, agriculture (greenhouses), aquaculture and tourism health. Hot springs and hammam righa Hammam Bouira (a) (b) Ksénna have a significant amount of geothermal energy with temperatures respectively of 64° C and 63° C and a flow rate of 20 l/s each. The use of this renewable energy would achieve energy savings of sizes in addition to contributing to the reduction of greenhouse gases.

Acknowledgements

Author would like to thank the Renewable Energy Development Center (CDER) for all facilities given for the finalization of this study. With great pleasure we send our sincere thanks for the National Agency for Hydraulic Resources (ANRH) for data provided and address a special thank for Mr. K. Abdeladim from CDER for his helpful comments.

Author details

Mébrouk Benziada
CDER Centre de Développement des Energies Renouvelables, Algiers, Algeria

*Address all correspondence to: mbenziada2011@gmail.com

References

[1] Adjoul – Dib H. Le thermalisme de l'Est algérien. Doctoral thesis, 3rd cycle – U.S.T.A. 287 p. 1984.

[2] Durand Delga M. Mise au point sur la structure du Nord-est de la Berbèrie. Publ. Serv. Géol. Algérie. Nlle Sér, 39, 89-131. 2007.

[3] Bouaicha, F. le géothermalisme de la région de Guelma. Doctorat en Sciences. Département des sciences géologiques. Faculté des sciences de la terre de la géographie et de l'aménagement du territoire. Université des sciences de Constantine , 2018.

[4] Fournier RO, and Trusdell AM. An empirical Na K Ca - Geothermometer for natural waters. Geochimica et Cosmochimica acta vol 37 pp. 1255-1275. 1973.

[5] Fournier RO. Chemical geothermometers and mixing model for geothermal systems.Geothermics, 5, 41-50. 1977.

[6] Gonord H, Grevellec J, Verdeil, P. Le thermalisme algérien dans son cadre géostructural. 5ème Réunion Annuelle des Sciences de la Terre – Rennes : 19 – 22, 1977.

[7] ENEL. Etude de reconnaissance géothermique Constantinois oriental. Report prepared for the National Society Electrical and Gaz Direction Renewables Technicals, Algiers, Algeria, 1982.

[8] Askri H, Belmecheri B, Boudjema A, Boumendjel K, Daoudi M, Drid M, Ghanem T, Docca A M, Ghandriche A, Ghomari A, Gellati N, Khennous, M, Lounici, R, Naili, H, Takherist D, Terkmani M. Geology of Algeria. Internal report Schlumberger-WEC, Sonatrach, 93 p. 1991

[9] Takherist D, Terkmani M. Geology of Algeria. Internal report Schlumberger-WEC, Sonatrach, 93 p. 1991.

[10] Athmania ,M . Mobilisation des ressources hydriques. Diplôme de Magistère Faculté des sciences de l'ingénieur Département de l'hydraulique appliquée. Université de Batna, 2006.

[11] Saibi H. Geothermal resources in Algeria. Renewable and Sustainable Energy Reviews 13, pp. 2544-3552, 2009.

[12] Bellache O, Hellel M, Abdelmalik EH, Chenak A. Chauffage de serres agricoles par énergie géothermique. CDER Rapport Interne 1984;12 [in French].

[13] Dickson M H, and Fanelli M. What Is Geothermal Energy, UNESCO Publication, Unterhaching, Germany, 30 May – 1 June 2007.

[14] Guigue S. Les sources thermo-minérales de l'Algérie. Etude géochimique (Tome II). 112 p. 1947.

[15] Guigue S. Les sources thermo-minérales de l'Algérie. Etude géochimique (Tome I). 140 p. 1940.

[16] Berkani, Ch ; Houha, B. Hydro-geochemical Characterization of Water Springs of the Aurès region. (North-easthern Algeria.J. Mater. Environ. Sci. 7 (6) (2016) 1856-1865. ISSN : 2028-2508 CODEN: JMESC.

[17] Site Web ADEME et BRGM, Géothermie – perspectives. http:/www. Géothermie – perspectives.fr. 2009.

[18] Touhami, F. Étude quantitative et qualitative des eaux thermales du Nord-Est algérien. TH7SE DE Doctorat en Sciences de l'eau Université des sciences de Montpellier, 2017.

[19] Boudoukha, A, Athmania, M , Benaabidat, L. Etude du potentiel hydrothermal de Hammam Sokhna Est algérien. Larhyss Journal, ISSN 1112-3680, n°24, Décembre 2015, pp. 161-174.

Chapter 7

Effect of Groundwater Flow and Thermal Conductivity on the Ground Source Heat Pump Performance at Bangkok and Hanoi: A Numerical Study

Arif Widiatmojo, Youhei Uchida and Isao Takashima

Abstract

In recent decades, the fast-growing economies of Southeast Asian countries have increased the regional energy demand per capita. The statistic indicates Southeast Asian electricity consumption grows for almost 6% annually, with space cooling becoming the fastest-growing share of electricity use. The ground source heat pump technology could be one of the solutions to improve energy efficiency. However, currently, there are limited data on how a ground source heat pump could perform in such a climate. The thermal response test is widely used to evaluate the apparent thermal conductivity of the soil surrounding the ground heat exchanger. In common practice, the apparent thermal conductivity can be calculated from the test result using an analytical solution of the infinite line source method. The main limitation of this method is the negligence of the physical effect of convective heat transfer due to groundwater flow. While convection and dispersion of heat are two distinctive phenomena, failure to account for both effects separately could lead to an error, especially in high groundwater flow. This chapter discusses the numerical evaluation of thermal response test results in Bangkok, Thailand, and Hanoi, Vietnam. We applied a moving infinite line source analytical model to evaluate the value of thermal conductivity and groundwater flow velocity. While determining the ground thermal properties in a high accuracy is difficult, the moving infinite line source method fulfills the limitation of the infinite line source method. Further, we evaluated the five-year performance of the ground source heat pump system coupled with two vertical ground heat exchangers in Bangkok and Hanoi. The results suggest the importance of groundwater flow to enhance the thermal performance of the system.

Keywords: Ground source heat pump, tropical region, Southeast Asia, space cooling, moving infinite line source, thermal response test

1. Introduction

The population of Southeast Asia was almost 640 million in 2016 and is expected to increase to 760 million by 2040, assuming 0.7% annual population growth.

Urbanization is an essential factor that affects total energy consumption [1–3]. The residential sector accounts for the second-highest electricity demand after the industrial sector, growing by an average annual rate of 7.5%, owing mainly to the increasing number of appliances. The introduction of energy-efficient products can restrain household energy demand. The Japan Refrigeration and Air Conditioning Industry Association (JRAIA) reported that in 2016, Vietnam and Thailand were the second and third countries with the highest air conditioner demand in Southeast Asia, with 1.98 million and 1.56 million units per year, respectively. Indonesia ranked first with 2.3 million units in the same year [4]. Annual regional air conditioner demand increased from 12.2 million units/year in 2011 to 16.4 million units/year in 2016, equivalent to an average of 6.1% increase per year.

Researches have been focused on the possibility of introducing the Ground Source Heat Pump (GSHP) in the region. Even though GSHP is a mature technology, the application of GSHP in the tropical climate, such as Southeast Asia, faces several problems. The use of GSHP is mainly for cooling, eliminating the balance between heat rejection and heat extraction. The ground temperature is relatively higher and within the range of air temperature.

Yasukawa et al. conducted underground temperature surveys by measuring vertical groundwater temperature from several monitoring wells in Thailand and Vietnam [5]. They concluded that despite the differences between ground temperature and air temperature were low. However, there are still advantages of utilizing GSHP for space cooling. Further, they remarked that space heating might be possible for a short winter period in Hanoi, Vietnam.

In their subsequent study, Yasukawa et al. presented the pilot study of GSHP at Kamphaengphet province, Thailand. They confirmed the applicability of the system with series of experimental tests [6].

Several studies have focused on providing further information on GSHP applicability in the regions. Widiatmojo et al. evaluated the performance of GSHP systems coupled with horizontal/shallow Ground Heat Exchanger (GHE). They also performed cost analysis to estimate the payback time against Air Source Heat Pump (ASHP). Shimada et al. examined the different operational conditions based on field experimentation and numerical simulation. While in another publication, Sasimook et al. presented the experiments and performance comparison of GSHP and ASHP. They highlighted the GSHP advantage, especially in higher thermal load [7, 8]. Although most of the studies above remarked the possibility of GSHP application in Southeast Asia, none of these addresses the effect of groundwater on the performance of GSHP.

This chapter discusses numerical simulation results to evaluate the Thermal Response Test (TRT) conducted in Bangkok, Vietnam, and Hanoi, Vietnam. The numerical simulation uses the Moving Infinite Line Source (MILS) analytical method to account for thermal conductivity and groundwater flow. Further, we extend the simulation to estimate the GSHP performance for five years of operation considering different parameters obtained using the Infinite Line Source (ILS) and the MILS methods.

2. Thermal response test

The Thermal Response Test (TRT) is a standard method to determine the ground thermal conductivity. From the TRT result, the apparent thermal conductivity of the ground surrounding the GHE can be calculated. A standard method to evaluate the apparent thermal conductivity from the TRT result is the Infinite Line Source (ILS) method [9, 10]. The ILS approach is based on the Kevin line source theory. This method calculates the temperature response of an infinite constant heat source analytically, assuming an infinite, isotropic, and homogeneous soil medium. This method also neglects the axial (vertical) heat transfer along the borehole. Considering the relationship between average fluid temperature, $\overline{T_f}$ (C) at a time t (s) with the borehole wall temperature at a radius r_b (m), constant heat–transfer rate per unit length of borehole, q (W m^{-1}) and borehole heat resistance, R_b (mK W^{-1}), the ILS solution is written as follows:

According to Carslaw and Jaeger [11], the temperature increase of a medium at a radial distance r (m) from an infinite line source with a constant heat exchange rate, q (Wm^{-1}), is expressed as:

$$T(r,t) - T_0 = \frac{q}{4\pi\lambda} E\left[\frac{r^2}{4Dt}\right] = \frac{q}{4\pi\lambda}\int_{\frac{r^2}{4Dt}}^{\infty} \frac{e^{-u}}{u} du \tag{1}$$

T_0 (C) is the initial ground temperature, D (m^2 s), the thermal diffusivity, and E is the exponential integral function. Assuming that the following condition is satisfied:

$$t > 5\frac{r^2}{D} \tag{2}$$

Eq. (1) can be re-written as:

$$\overline{T_f}(t) - T_0 = \left(T_f(r_b, t) + qR_b\right) - T_0 \cong qR_b + \frac{q}{4\pi\lambda}\left[ln\left(\frac{4Dt}{r_b^2}\right) - \gamma\right] \tag{3}$$

where, R_b (mK W^{-1}) is the borehole thermal resistance, r_b (m) is the borehole radius λ (W m^{-1} K^{-1}) is the thermal conductivity, D (m^2 s) is the thermal diffusivity, ϒ (−) is Euler constant, and T_0 (C) is the soil temperature at initial (t = 0). $\overline{T_f}$ (C) is the average circulation fluid temperature calculated by:

$$\overline{T_f} = \frac{(T_{bh-in} + T_{bh-out})}{2} \tag{4}$$

where, $T_{bh\text{-}in}$ and $T_{bh\text{-}out}$ (C) are GHE fluid inlet and outlet temperature, respectively. Eq. (3) can be re-arranged into the linear form of fluid temperature against the natural logarithmic value of time as:

$$\overline{T_f}(t) \cong \frac{q}{4\pi\lambda} ln(t) + \left(qR_{bh} + \frac{q}{4\pi\lambda}\left[ln\left(\frac{4Dt}{r_{bh}^2}\right) - \gamma\right] + T_0\right) = m\ ln(t) + c \tag{5}$$

From the fluid temperature gradient against the natural logarithmic value of time, m (C), obtained from TRT measurement, the value of apparent thermal conductivity, λ_{app} (W m^{-1} K^{-1}) can be calculated as:

$$\lambda_{app} = \frac{q}{4\pi m} \tag{6}$$

In the ILS method, the effect of convective heat transfer as a result of groundwater convection is not considered. The value of apparent thermal conductivity represents both diffusive and convective heat transfer. Accordingly, it is recognized that the value of apparent thermal conductivity is larger than the value of effective thermal conductivity, λ_{eff}. The heat transfer due to convection of groundwater flow and heat conduction are two different physical phenomena. Thus, the use of apparent thermal conductivity (heat conduction) for calculating the thermal performance of vertical ground heat exchangers can lead to some serious errors, especially for the longer time-scale and high-velocity groundwater flow.

While groundwater flow is an important parameter, measuring the groundwater velocity is practically difficult. Besides, the ground layers are inhomogeneous. The practical way to measure the groundwater velocity is the pumping test. However, the pumping test is expensive as it requires an additional borehole for the observation well.

The TRT measurements were carried out in Bangkok, Thailand and Hanoi, Vietnam. These were the first and second measurements to be carried out in Southeast Asia [12]. The measurements were conducted in the existing GSHP systems. In Bangkok, TRT measurement was performed in an installed GSHP system at Chulalongkorn University, Bangkok campus. The measurement was also performed in the GSHP system installed at the Vietnam Institute of Geosciences and Mineral Resources (VIGMR), Hanoi. The measurement procedures were similar for both sites. **Figure 1** shows the measurements at Bangkok and Hanoi.

Both measurements applied the constant heating rates q = 39.72 Wm^{-1} and q = 35.91 Wm^{-1} for Bangkok and Hanoi, respectively. By evaluating the TRT results using the ILS method, the apparent thermal conductivity was calculated as λ_{app} = 1.82 W m^{-1} K^{-1} and λ_{app} = 1.42 W m^{-1} K^{-1} for Bangkok and Hanoi, respectively. The data regarding the effective thermal conductivities and groundwater velocities in both GSHP sites are unavailable. Further details on the measurements can be found in another publication [12].

Figure 1.
TRT measurement at the GSHP sites: Chulalongkorn University, Bangkok (left) and VIGMR, Hanoi (right).

3. Moving infinite line source

To consider the effect of groundwater flow, we evaluated the TRT measurement results in Bangkok and Hanoi by applying the Moving Infinite Line Source (MILS) theory. According to Diao et al., the temperature increase at a radial position, φ(rad), from a line source is expressed as [13]:

$$T(r, \varphi, t) - T_0 = \frac{q}{4\pi\lambda} \exp\left(\frac{u_{eff} r}{2D} \cos\varphi\right) \int_0^{r^2/4Dt} \frac{1}{\eta} \exp\left[-\frac{1}{\eta} - \frac{u_{eff}^2 r^2 \eta}{16D^2}\right] d\eta \tag{7}$$

where, $\eta = 4D(t\text{–}t')/r^2$, u_{eff} (m^2s^{-1}) is the effective velocity of groundwater flow assuming local thermal equilibrium, calculated by:

$$u_{eff} = u\frac{\rho_w c_w}{\rho c} \quad (8)$$

Here, u (ms^{-1}) is the seepage velocity, ρ_w (kgm^{-3}) and c_w ($Jkg^{-1}K^{-1}$) are the volumetric mass density and the specific heat of water, respectively. The following relationship defines the volumetric mass density and specific heat of the medium:

$$\rho c = (1-\varepsilon)\rho_s c_s + \varepsilon\rho_w c_w \quad (9)$$

Where ρ_s (kgm^{-3}) and c_s ($Jkg^{-1}K^{-1}$) are the volumetric mass density and the specific heat of the soil matrix, respectively. Eq. (7) calculates the temperature of the soil medium at an arbitrary position adjacent to the line source. The following equation represents heat balance between average fluid temperature and borehole wall temperature [14]:

$$\overline{T}_f(t) = \frac{1}{2\pi}\int_0^{2\pi} T(r_{bh}, \varphi, t)\mathrm{d}\varphi + qR_{bh} \quad (10)$$

where R_{bh} (mKW^{-1}) is the borehole heat resistance.

The simulation using the MILS analytical solution is valid under the following assumptions:

- The soil medium homogenous and infinite
- The ground physical and thermal properties are independent of time and temperature
- The effect of ambient temperature and the boundary between ground and surface are negligible
- The initial ground temperature is uniform

4. Discussion

Figure 2 shows the TRT measurement results and the numerical simulation using Eq. (7) by setting the u_{eff} = 0 and apparent thermal conductivity similar to those calculated using the ILS method. Additional parameters for the numerical simulations are listed in **Table 1**. The discrepancies between simulations and measurements at the beginning are likely the indication of heat transfer within boreholes, which have different thermal properties than the surrounding soil [14]. It is essential to ensure the apparent thermal conductivity adequately represents the value of soil mass. Typically, only the last few hours of results are considered for the linear fitting of Eq. (5).

4.1 Estimation groundwater flow velocity and effective thermal conductivity

By using the MILS method, the groundwater flow can be taken into account. However, the determination of groundwater velocity remains a problem. A

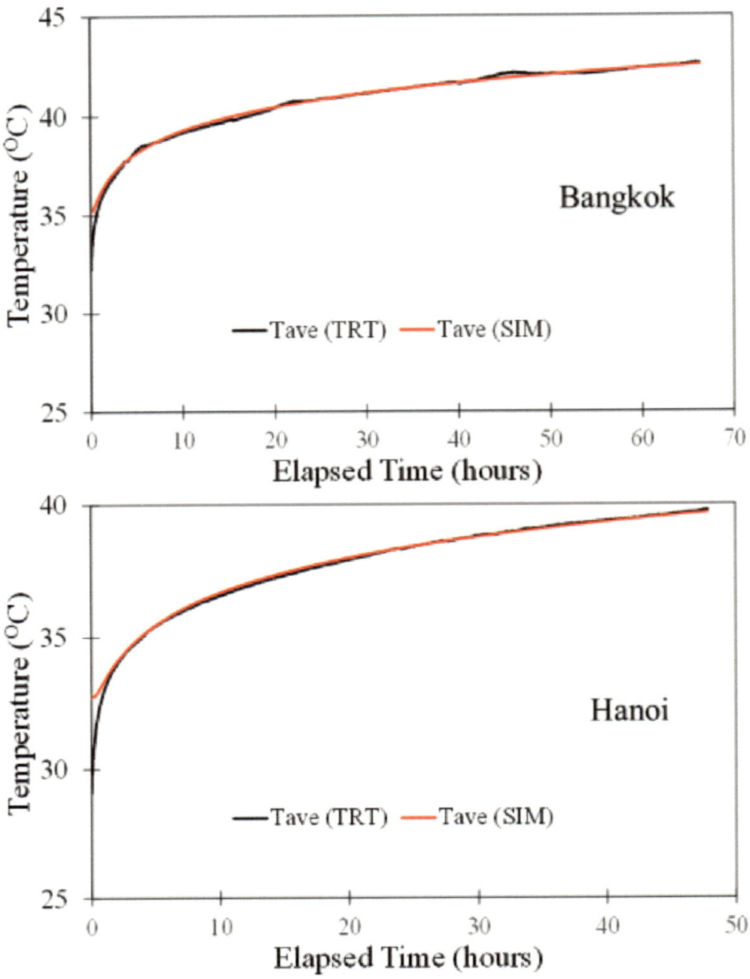

Figure 2.
The average fluid temperatures obtained from TRT measurements and the simulation results using Eq. (7).

	T_0 (°C)	R_{bh} (mK W^{-1})	q (Wm^{-1})	λ (Wm^{-1} K^{-1})	D (m^{2}s)
Bangkok	29	0.156	39.72	1.82	7.58e-7
Hanoi	27	0.16	35.91	1.42	5.07e-7

Table 1.
Simulation parameters for TRT data using Eq. (7) (u_{eff} = 0).

numerical approach incorporating a parameter optimization method has been proposed [14, 15]. In this study, a similar numerical procedure was performed in Matlab employing the *fminsearch* function. The *fminsearch* is a pre-programmed function to search the minimum unconstrained multivariable function using the derivative-free method. The parameter estimation from the TRT results employs the *fminsearch* function to find the minimum value of Root Mean Square Error (RMSE) between the MILS model and the TRT result [14].

$$RMSE = \sqrt{\frac{\sum_{i=1}^{N}\left(\overline{T_{f(sim)}}(i) - \overline{T_{f(meas)}}(i)\right)^2}{N}} \tag{11}$$

The initial values of the fitting parameters are given; thermal conductivity, effective groundwater velocity, and borehole heat resistance. Three values are provided for each parameter, resulting in a total

$$\overline{T}_{f(sim)}$$

of 27 combinations. Some of the results from different combinations yield similar results. To summarize the results, we select some representative values from these combinations, as shown in **Table 2**.

Figure 3 compares TRT results and the simulation results considering the optimized parameters listed in **Table 2** for both sites. The simulation results fit well with the TRT measurements. The higher RMSE error for the Bangkok site is due to the fluctuations of the measured data. As expected, the MILS models predict smaller thermal conductivities than those calculated using the ILS method for all cases. The smaller thermal conductivities are because the MILS models account for the effect of groundwater flow. It is interesting to note that the H1, H2 and H3 yield a similar value of thermal conductivity and RMSE but differs in groundwater velocities. The predicted velocities are low, and their effect on the calculated temperature response is not significant. In addition, each model shows the convergence of borehole thermal resistance values, except for the case of B1. Overall, the differences between the simulated temperatures are close to each other and visually difficult to be distinguished.

4.2 Effect of groundwater flow and ground thermal conductivity on the performance of ground heat exchanger

In the previous discussion, the effect of different thermal conductivities and groundwater velocities over the short TRT measurement period is not distinct. To further examine the impact of these parameters on the fluid temperatures, we extend the simulation period to one year. The simulations assume constant heating rates similar to the field TRT measurement. The simulation results are shown in **Figures 4** and **5** for Bangkok and Hanoi models, respectively. The calculations considering the thermal conductivity values obtained using the ILS method are also presented. For the Bangkok cases, the discrepancies due to the various estimated thermal conductivities and effective groundwater velocities are clearly observable. The B3 and B4 cases, which have the lowest predicted effective groundwater velocities, show higher average fluid temperatures. Meanwhile, B1 and B2 cases yield lower average fluid temperatures.

Interestingly, the average fluid temperature converges into an asymptotic value for the highest effective groundwater velocity (B1 case) after few days. It reveals the critical role of the convective heat transfer to the ground heat exchanger performance over an extended period.

		λ_{fit}	u_{eff}	R_{bh}	RMSE
Bangkok	B1	1.45	1.82E-06	0.1388	0.0706
	B2	1.68	4.55E-07	0.1486	0.0770
	B3	1.69	1.01E-07	0.1493	0.0777
	B4	1.69	1.37E-07	0.1493	0.0776
Hanoi	H1	1.34	2.88E-10	0.1537	0.0258
	H2	1.34	2.96E-09	0.1537	0.0258
	H3	1.34	1.14E-07	0.1536	0.0258
	H4	1.34	1.14E-08	0.1537	0.0258

Table 2.
List of parameters from the MILS fitting simulation against the TRT data.

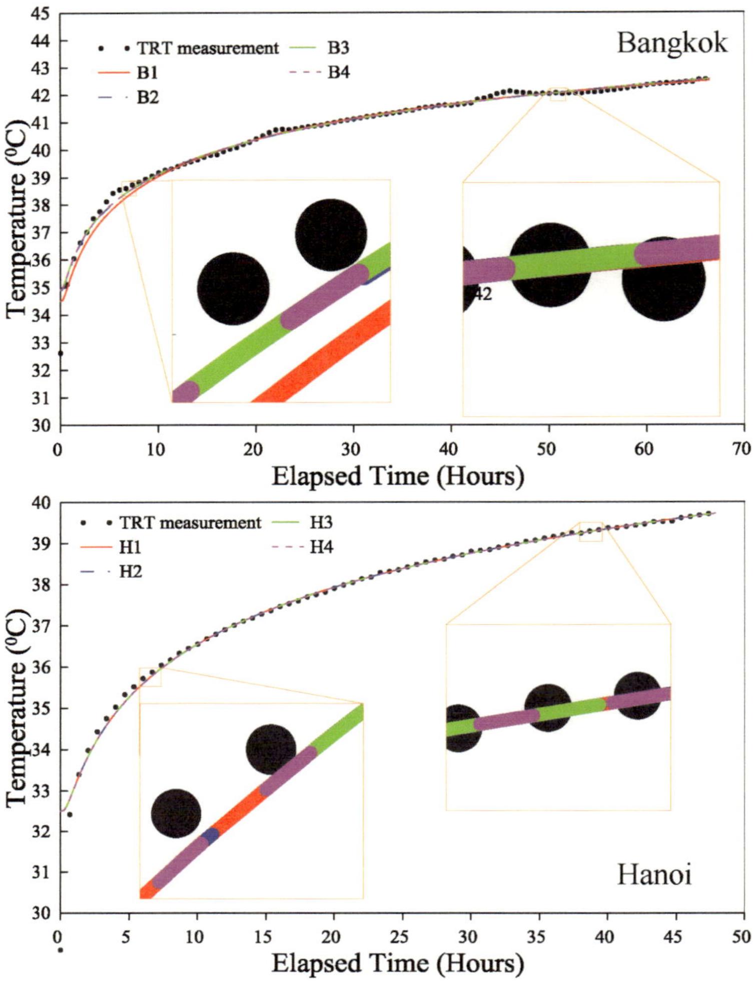

Figure 3.
*The average fluid temperatures from TRT measurements and the best fits MILS simulation using parameters as listed in **Table** 2.*

Figure 5 shows the differences over the extended period for the Hanoi numerical model. While H1, H2, and H4 yield almost similar values of average fluid temperatures, the highest predicted effective groundwater velocity, the H3 case, shows the lower average fluid temperature.

Figures 6 and 7 show the contour plots representing the soil temperature increase after a year of constant heating. The ground heat exchanger is located at the center coordinate (0,0), and the grid intervals are shown in meter-unit. The groundwater flows to the positive x-direction (in **Figure 6 right** and **Figure** 7 **right**). The left-hand part in both **Figures 6** and 7 is the simulation result in case the apparent thermal conductivity calculated by the ILS method is used (u_{eff} = 0). In comparison, the right-hand part is the simulation result using the optimized value of thermal conductivity and groundwater velocity (B2 and H3 cases). The maximum temperature and the shape of isothermal lines for Bangkok show that the case with groundwater velocity u_{eff} = 4.55E-07 ms^{-1}, despite its lower thermal conductivity, provides better thermal performance than the case with u_{eff} = 0. On the contrary, the numerical results for Hanoi cases (**Figure** 7) show the opposite due to the low groundwater velocity.

Despite the advantage of MILS, the reverse analysis involving parameter optimization performed in this study results in several combinations of parameters. The

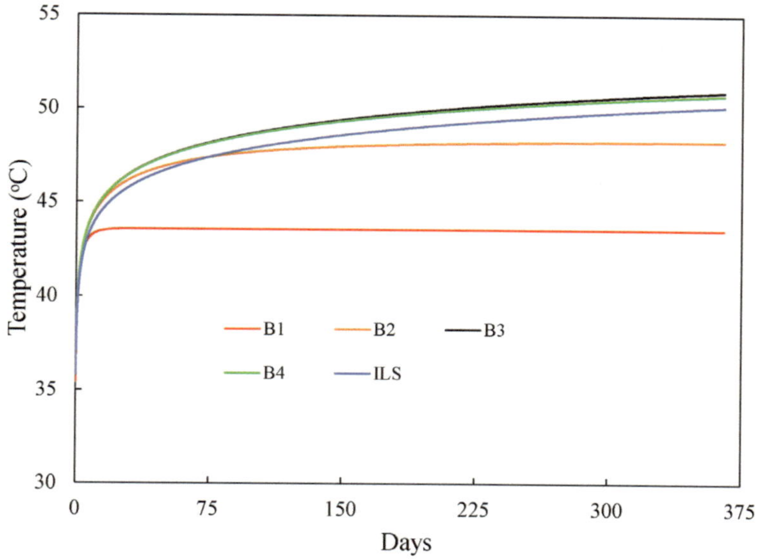

Figure 4.
Effect of different parameters (thermal conductivity and effective groundwater velocity) to a year cycle of average fluid temperature under a constant heat rejection rate (q = 39.72 Wm⁻¹) for Bangkok case.

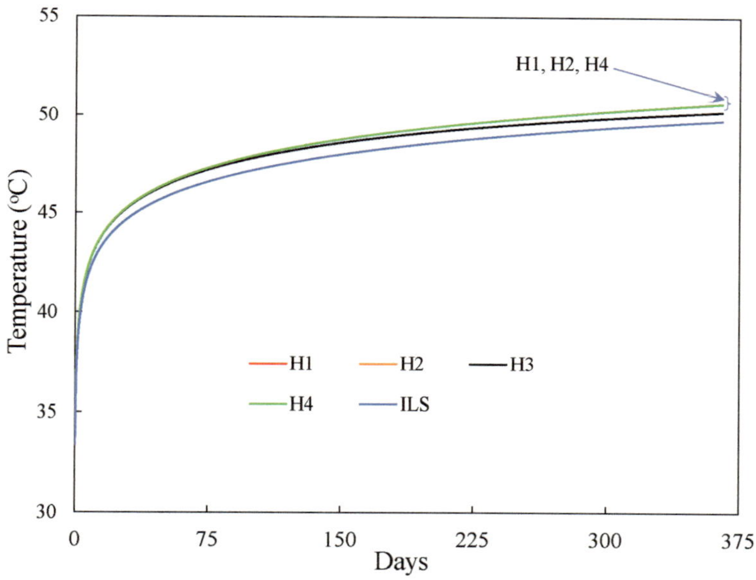

Figure 5.
Effect of different parameters (thermal conductivity and effective groundwater velocity) to a year cycle of average fluid temperature under a constant heat rejection rate (q = 35.91 Wm⁻¹) for the Hanoi case.

improvement should be emphasized on the TRT method to provide data to narrow down the resulting combinations into the best possible solutions. One of the most feasible methods is providing different heating rates for a TRT site.

4.3 Long-term GSHP performance in Bangkok and Hanoi

One of our main interests in the GSHP potential application in Southeast Asia is to estimate the GSHPs long-term sustainability. Thus, it is essential to examine further the effect of different ground thermal properties on the long-term

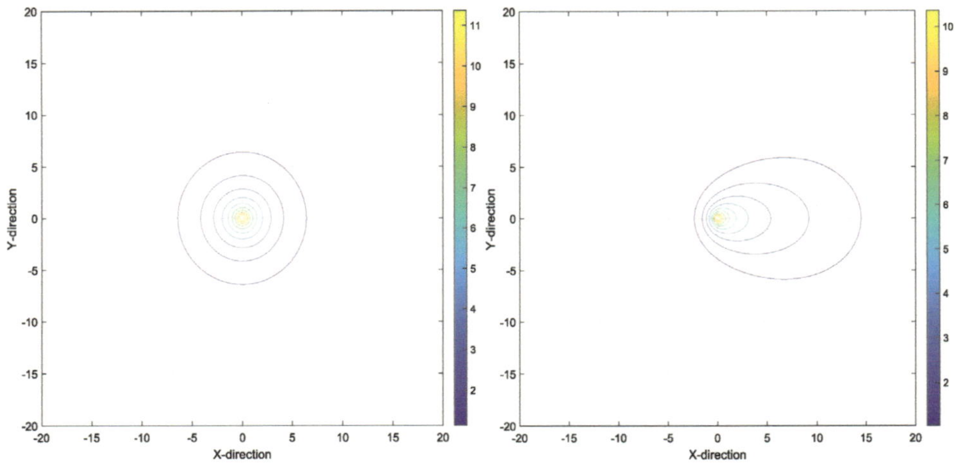

Figure 6.
Contour plot showing the soil temperature increase after a year of constant heating for the Bangkok site; (left: Q = 39.72 Wm^{-1}, λ = 1.82 $Wm^{-1}K^{-1}$, u_{eff} = 0 ms^{-1}; right: Q = 39.72 Wm^{-1}, λ = 1.68 $Wm^{-1}K^{-1}$, u_{eff} = 4.55E-07 ms^{-1}).

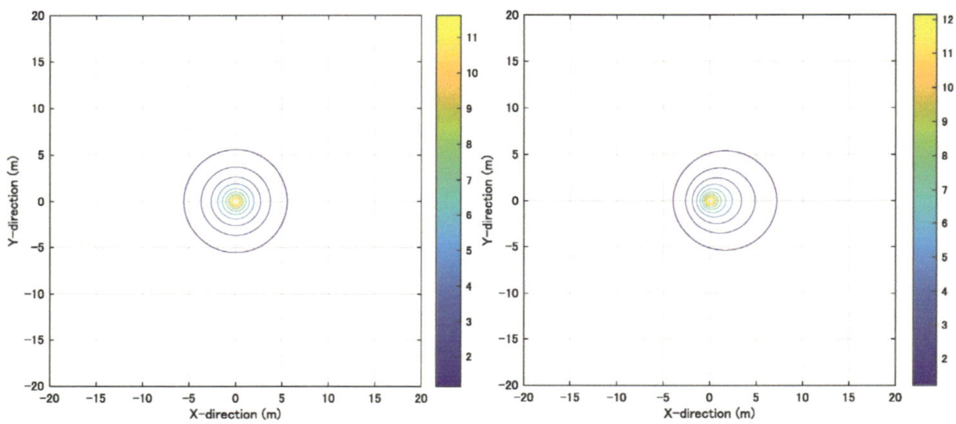

Figure 7.
Contour plot showing the soil temperature increase after a year of constant heating for the Hanoi site; (left: Q = 35.91 Wm^{-1}, λ = 1.42 $Wm^{-1}K^{-1}$, u_{eff} = 0; right: Q = 35.91 Wm^{-1}, λ = 1.34 $Wm^{-1}K^{-1}$, u_{eff} = 1.14E-07 ms^{-1}).

performance of GSHPs. So far, the MILS analytical model is only limited to a single ground heat exchanger. Here, we propose a simple modification to the described analytical models to simulate two ground heat exchangers.

The maximum number of ground heat exchangers in the numerical model depends on the symmetrical arrangement of boreholes and whether the model considers the groundwater flow. The simplest numerical model disregarding the effect of groundwater flow can simulate up to four ground heat exchangers. The numerical model can simulate a maximum of two ground heat exchangers if the groundwater flow is considered. In such a case, the boreholes must be arranged perpendicular to the groundwater flow direction.

The calculation of the temperature field uses the superposition method. Once the average fluid temperature for one of the boreholes is calculated, the total average fluid temperature flowing from/to the heat pump can be calculated owing to its symmetrical arrangement. The following additional equation is required to calculate the unknown inlet and outlet temperatures, $T_{bh\text{-}in}$ and $T_{bh\text{-}out}$.

$$T_{bh-in} - T_{bh-out} = Q/_{\rho_f \vartheta c_f} \tag{12}$$

Where Q (Watt) is the heat rejection rate from heat-pump to the ground, ρ_f (kgm^{-3}), ϑ (m^3s^{-1}) is the volumetric flowrate of heat exchanger fluid, and c_f ($Jkg^{-1}K^{-1}$) is the specific heat capacity of heat exchanger fluid. The inlet and outlet temperatures can be calculated by combining Eqs. (4) and (12).

The long-term simulation model considers a GSHP system with a heat rejection rate of 5 kW. The heat pump connects to two vertical 50 m-long ground heat exchangers in a parallel flow configuration. A parallel flow configuration means that the heat exchanger fluid flows from the heat pump into each borehole at a proportional flow rate (see **Figure 8**). Thus, the heat exchange rate per unit length is identical for both ground heat exchangers. In the present study, the GSHP is assumed to be used only for cooling purposes (heat rejection) and operates only 8 hours a day (8 am to 4 pm) during weekdays. These assumptions are to represent the typical behavior of air conditioner use in standard office buildings. In addition, the simulation period is five years. The numerical models disregard the effect of ambient air temperature fluctuations on the cooling load.

A standard parameter to evaluate the thermal performance of GSHP is the Coefficient of Performance (COP). The COP is a ratio between the total rate of cooling or heating to the required electrical input. While the COP is affected by various factors, a simple approximation is possible via a correlation with the heat pump's fluid temperature (T_{bh-out}) [16]. Such correlation can be obtained from the performance tables provided by the manufacturer. **Figure 9** shows the correlation between COP and T_{bh-out} for a heat pump with a rated capacity of 5.27 kW. This correlation is specific for a 15.8 L min^{-1} fluid flowrate and 27°C dry-bulb and 19°C room air temperature at the 13.45 m^3min^{-1} flow rate (air flowrate in the fan-coil unit).

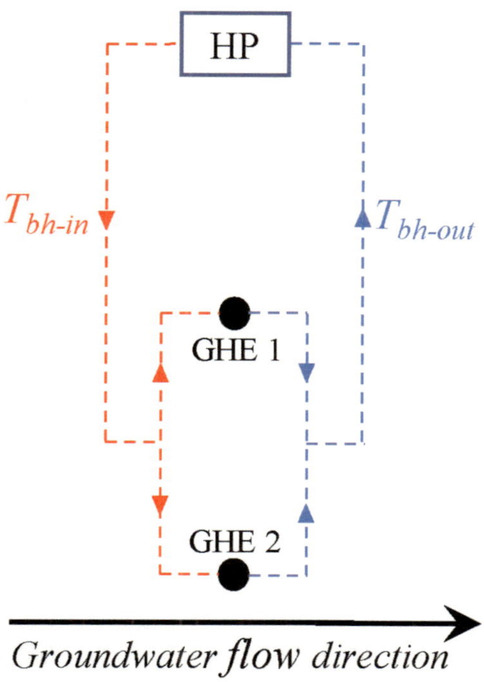

Figure 8.
Schematic figure of the numerical model for long-term performance evaluation.

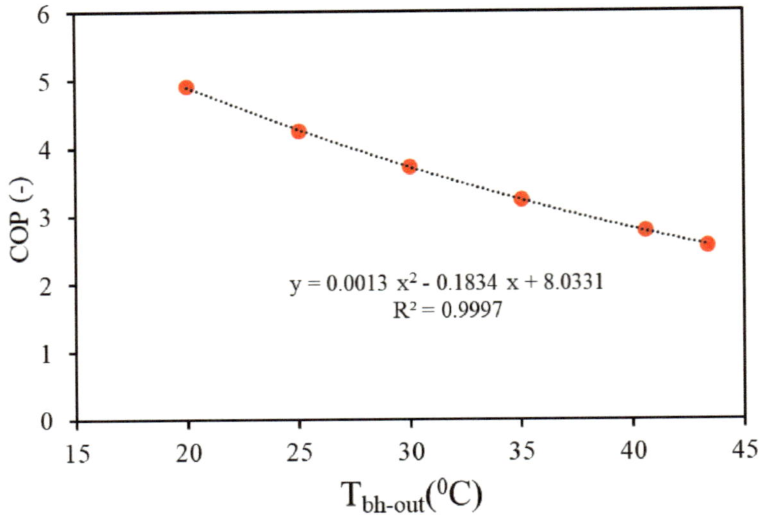

Figure 9.
A relationship between $T_{bh\text{-}out}$ *and COP of a heat pump, obtained from the performance table supplied by the manufacturer.*

Figures 10 and **11** present the fluid temperature leaving the ground heat exchanger ($T_{bh\text{-}out}$) and the calculated COP, respectively. The initial ground temperature and ground thermal properties are similar to those applied in the previous simulations (see **Tables 1** and **2**). The simulation results considering the thermal properties calculated using the ILS method show higher fluid temperatures than the simulations with the groundwater flow (case B4 for Bangkok and H3 for Hanoi). At the end of the five-year operation, the final fluid temperature for Bangkok are 41.77°C and 41.1°C for ILS and B4 cases, respectively. While, the final fluid temperatures for Hanoi are 41.89°C and 40.86°C for ILS and H3 case, respectively. Note that the thermal conductivities for B4 and H3 cases are lower than those calculated using the ILS. **Figure 12** compares the iso-temperature plot of B4 and H3 cases after five years. The B4 case with higher thermal conductivity and groundwater velocity provides a better heat transfer rate.

The results also suggest the significant role of groundwater convection in lowering the fluid temperature. For the application of GSHP in tropical countries with high initial ground temperature, the role of groundwater flow is ultimately essential.

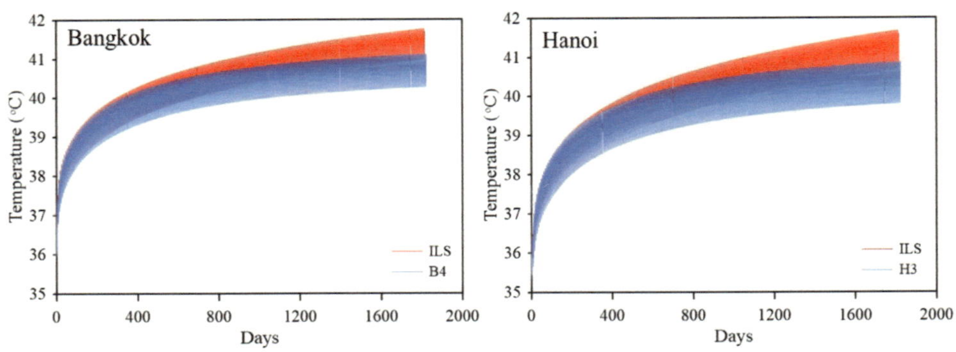

Figure 10.
Fluid temperature flowing from the ground heat exchanger ($T_{bh\text{-}out}$) *for five-years GSHP operational period.*

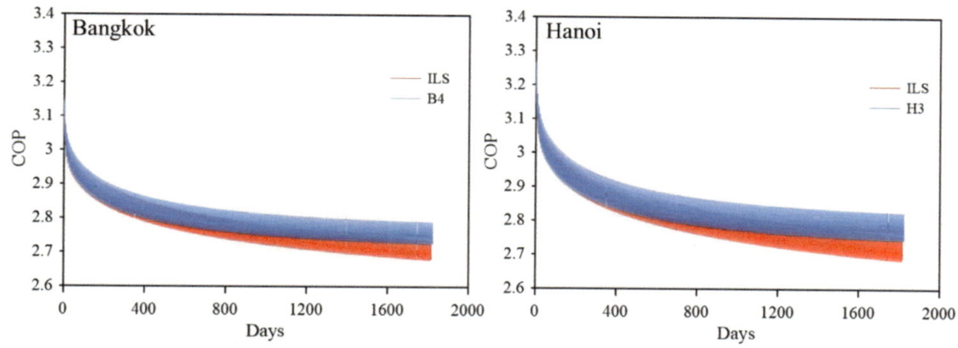

Figure 11.
COP of the GSHP system for a five-year GSHP operational period.

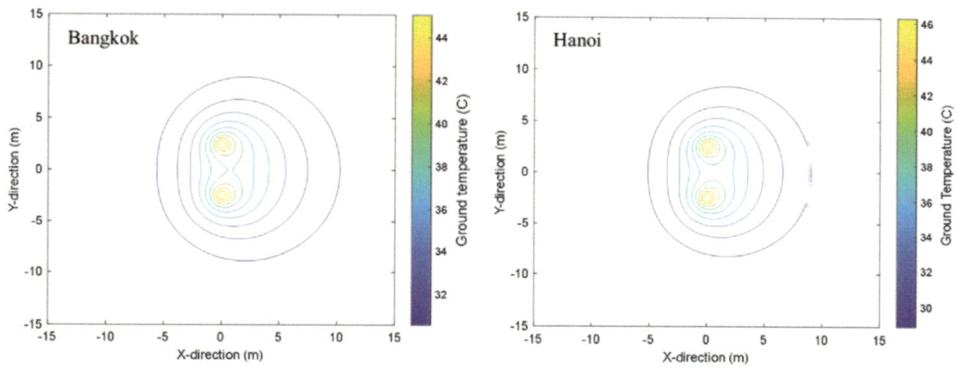

Figure 12.
Contour plot showing the soil temperature increase after a five-year GSHP operation; the thermal parameters for B4 and H3 cases are used for Bangkok (left) and Hanoi (right), respectively.

5. Conclusion

The application of GSHP in tropical countries, such as Thailand and Vietnam, encounters several problems. One of the main problems is the insufficient data on how GSHP could perform under such a climate. We carried out the numerical analysis of the TRT measurement results conducted in Bangkok, Thailand and Hanoi, Vietnam.

The MILS analytical method provides a better numerical analysis to evaluate the TRT result than the commonly used ILS method. The inverse analysis of TRT result using the MILS method with parameter optimization resulted in multiple solutions of unknown parameters: groundwater flow velocity, thermal conductivity, and borehole thermal resistance. The groundwater velocity and thermal conductivity are parameters with significant variations, while the borehole thermal resistance indicates a stable convergence into a single value. Simulations considering parameters obtained using the MILS and ILS method do not show a clear difference over a short-term TRT period. In the extended period of simulations, more than a year time scale, the differences are evident. The results also suggest the importance of groundwater flow in the long-term performance of GSHP, especially in tropical regions with high soil background temperatures. Over-reliance on the ILS method and the use of apparent thermal conductivity, especially for the site with high groundwater flow, can lead to a severe error.

Further, we extended the MILS simulations by incorporating two borehole heat exchangers and a simple approximation of COP to evaluate the GSHP performance over five years under different ground thermal parameters. The COP decreases over a more extended period because of the thermal imbalance resulting from the absence of heat extraction. The simulation results also suggest that the groundwater flow can effectively reduce the decreasing rate of COP.

Acknowledgements

The authors would like to thank Prof. Hikari Fujii and Mr. Hiroyuki Kosukegawa of Akita University for the helpful discussion and support during TRT measurements. Also, the authors wish to extend special thanks to Prof. Punya Charusiri, Dr. Srilert Chotpantarat, Ms. Sasimook Chokchai of Chulalongkorn University, Thailand, and also Mr. Trong Thang Tran of Vietnam Institute of Geoscience and Mineral Resources, Vietnam, for their valuable support in this research.

This research is partially supported by:

- The Leading Initiative for Excellent Young Researcher (LEADER), Ministry of Education, Culture, Sport, Science and Technology, Japan.

- Coordinating Committee for Geoscience Program in East and Southeast Asia (CCOP), Groundwater Sub-Project.

Author details

Arif Widiatmojo[1*], Youhei Uchida[1] and Isao Takashima[2]

1 Renewable Energy Research Center, National Institute of Advanced Industrial Science and Technology, Koriyama-shi, Fukushima-ken, Japan

2 Mining Museum, Akita University, Akita-shi, Akita-ken, Japan

*Address all correspondence to: arif.widiatmojo@aist.go.jp

References

[1] Le VT, Pitts A. A survey on electrical appliance use and energy consumption in Vietnamese households: Case study of Tuy Hoa city. Energy Build. 2019;197: 229–241. DOI: 10.1016/j.enbuild.2019.05.051

[2] Bakirtas T, Akpolat AG. The relationship between energy consumption, urbanization, and economic growth in new emerging-market countries. Energy. 2018;147:110–121. DOI: 10.1016/j.energy.2018.01.011

[3] Doan VQ, Kusaka H. Projections of urban climate in the 2050s in a fast-growing city in Southeast Asia: The greater Ho Chi Minh City metropolitan area, Vietnam. Int J Climatol. 2018;38 (11):4155–4171.

[4] Japan Refrigeration and Air Conditioning Industry Association (JRAIA). World Air Conditioner Demand by Region. 2018.

[5] Yasukawa K, Uchida Y, Tenma N, Taguchi Y, Muraoka H, Ishii T, et al. Groundwater Temperature Survey for Geothermal Heat Pump Application in Tropical Asia. Bull Geol Surv Japan. 2009;60(9–10):459–467. DOI: 10.9795/bullgsj.60.459

[6] Yasukawa K, Takashima I, Uchida Y. Geothermal heat pump application for space cooling in Kamphaengphet , Thailand. Bull Geol Surv Japan. 2009;60 (9/10):491–501.

[7] Shimada Y, Uchida Y, Takashima I, Chotpantarat S, Widiatmojo A, Chokchai S, et al. A study on the operational condition of a ground source heat pump in Bangkok based on a field experiment and simulation. Energies. 2020;13(1).

[8] Chokchai S, Chotpantarat S, Takashima I, Uchida Y, Widiatmojo A, Yasukawa K, et al. A Pilot Study on Geothermal Heat Pump (GHP) Use for Cooling Operations, and on GHP Site Selection in Tropical Regions Based on a Case Study in Thailand. Energies. 2018; 11(9):2356. DOI: 10.3390/en11092356

[9] Choi W, Ooka R. Effect of disturbance on thermal response test, part 2: Numerical study of applicability and limitation of infinite line source model for interpretation under disturbance from outdoor environment. Renew Energy. 2016;85:1090–1105. DOI: 10.1016/j.renene.2015.07.049

[10] Signorelli S, Bassetti S, Pahud D, Kohl T. Numerical evaluation of thermal response tests. Geothermics. 2007;36 (2):141–166.

[11] Carslaw HS, Jaeger JC. Conduction of Heat in Solids. Newyork: Oxford University Press; 1959.

[12] Widiatmojo A, Shimada Y, Takashima I, Uchida Y, Chotpantarat S, Charusiri P, et al. Evaluation and Short-Term Test on Potential Utilization of Ground Source Heat Pump for Space Cooling in Southeast Asia. In: Hazarika, H., Madabhushi GSP, Yasuhara K, Bergado DT, editors. Advances in Sustainable Construction and Resource Management. Springer Singapore; 2021. DOI: 10.1007/978-981-16-0077-7

[13] Diao N, Li Q, Fang Z. Heat transfer in ground heat exchangers with groundwater advection. Int J Therm Sci. 2004;43(12):1203–1211.

[14] Angelotti A, Ly F, Zille A. On the applicability of the moving line source theory to thermal response test under groundwater flow: considerations from real case studies. Geotherm Energy. 2018;6(1). DOI: 10.1186/s40517-018-0098-z

[15] Ly F. Interpretation of Borehole Heat Exchangers Thermal Response

Tests under groundwater influence: analysis of three case studies. Master Degree Thesis. 2015.

[16] Nam Y, Ooka R, Hwang S. Development of a numerical model to predict heat exchange rates for a ground-source heat pump system. Energy Build. 2008;40(12):2133–2140.